COMMUNITY-BASED MONITORING INITIATIVES OF WATER AND ENVIRONMENT:

EVALUATION OF ESTABLISHMENT DYNAMICS AND RESULTS

T0176368

Mohammad Gharesifard

COMMUNITY-BASED MONITORING INITIATIVES OF WATER AND ENVIRONMENT: EVALUATION OF THE ESTABLISHMENT DYNAMICS AND RESULTS

DISSERTATION

Submitted in fulfillment of the requirements of

the Board for Doctorates of Delft University of Technology

and

of the Academic Board of the IHE Delft

Institute for Water Education

for

the Degree of DOCTOR

to be defended in public on

Monday, 28 September 2020, at 10:00 hours

in Delft, the Netherlands

by

Mohammad GHARESIFARD

Master of Science in Water Resources Management

IHE Delft Institute for Water Education, Delft

Born in Shiraz, Iran

This dissertation has been approved by

Prof. dr. ir. P. van der Zaag   IHE Delft and TU Delft, promotor
Dr. U. Wehn   IHE Delft, copromotor

Composition of the doctoral committee:

Rector Magnificus TU Delft   Chairman
Rector IHE Delft   Vice-Chairman

Independent members:

Prof. dr. ir. N.C. van de Giesen   TU Delft
Prof. dr. ir. C. Leeuwis   Wageningen University
Prof. dr. rer. nat. A. Bonn   German Centre for Integrative Biodiversity Research
Dr. D. Kasperowski   University of Gothenburg
Prof. dr. M.E. McClain   IHE Delft and TU Delft, reserve member

This research was conducted under the auspices of the SENSE Research School for Socio-Economic and Natural Sciences of the Environment

Published by:
CRC Press/Balkema
Schipholweg 107C, 2316 XC, Leiden, the Netherlands
Pub.NL@taylorandfrancis.com
www.crcpress.com – www.taylorandfrancis.com
ISBN: 978-0-367-67401-4

To Bahar, whose never ending love, patience and support made this journey possible

ACKNOWLEDGMENTS

This is the last section that I wrote in this dissertation and arguably one of the most difficult ones. Looking back at a journey of almost five years, I realized how much support, love and encouragement I have received from so many people around me. I did my best to be inclusive and mention everyone who has contributed to the content of this work or have provided me with supported for completing this work in one way or another.

This research was funded by the WeSenseIt and Ground Truth 2.0 projects that received funding from the European Union's Seventh Framework Programme for Research and Technological Development (FP7) and Horizon 2020, respectively.

First and foremost, I would like to express my deepest gratitude to my promoter Professor Pieter van der Zaag and my co-promoter Dr. Uta Wehn who have substantially contributed to the work presented in this dissertation and also to my professional and personal growth during the past five years. Pieter, since my master's studies, you have been an exemplary teacher and mentor for me. Your critical and constructive views and comments have helped me a lot with looking at the details of my research from a different angle and rethinking the meaning of my findings. Your critical views, ethics in academia and an almost fatherly like attention to your students have made you my role model in academia and beyond. Uta, I am immensely grateful for the trust that you have put in me and for the extensive inputs that you have provided on this research. Without your trust, continued support and inputs, this research would not have been possible. I will never forget your critical, thorough and sometime overwhelming comments on my rather mechanical writing style. These comments challenged my inner engineer, pushed me to my limits and helped me to expand those limits to quite to a large extent.

I would like to express my gratitude to IHE Delft, Institute for Water Education and TU-Delft that were my host institutes and provided me with organizational support during my PhD studies. I am especially thankful for the support that I have received from Jolanda Boots, Mariëlle van Erven, Floor Felix, Anique Karsten at IHE Delft and from Lydia de Hoog at the Water Management Department of TU-Delft. I also would like to offer my special thanks to colleagues and staff members at IHE Delft, with whom I interacted during my PhD, whose words have encouraged me, or whose works have inspired me during this journey. Thank you so much Maria Rusca, Jeltsje Kemerink, Margreet Zwarteveen, Klaas Schwartz, Ilyas Masih, Janez Susnik, Frank Jaspers, Leonardo Alfonso, Hermen Smit, Mireia Tutusaus, Zaki Shubber, Ioana Popescu, Guy Alaerts, Andreja Jonoski, Claire Michailovsky, and Yong Jiang.

I consider myself very lucky to have been able to conduct this research in the context of two major EU-funded projects. During my PhD, I had the pleasure of interacting and working with quite a few partner organizations of the WeSenseIt and the Ground Truth 2.0 projects. Because of the selection of the case studies of this research, I interacted most with staff members of my home organizations IHE Delft, as well as TAHMO, Upande and HydroLogic Research. Dear colleagues Kim Anema and Hans van der Kwast, thank you both for all your support and for helping me better understand the context of the Dutch and the Kenyan case studies. Kim, I really appreciate you sharing your insights after the late evening stakeholder meetings in the Dutch case, while we were driving back from Altena to Delft. Hans, I will never forget my first trip to Kenya, in which you tried to explain every single detail about the study area and you introducing me to the beautiful Maasai culture. I am particularly thankful to Frank Annor from TAHMO for the lively and fruitful discussions that we had during my visits to Kenya. Mark de Blois, Luchiri Omoto, Simon Gathuru and other colleagues at Upande; thank you all for your help and support for conducting this research in Kenya. I deeply appreciate Rianne Giesen, Annemarleen Kersbergen, Christian Slijngard, Marlies Zantvoort, Francine Teeuwen, Sanne Veldhuijsen and Elsje Burgers from HydroLogic Research, who facilitated, conducted or translated a number of interviews in the Dutch case study of this research.

This research is built on opinions, local knowledge and inputs of 92 interviewees in Kenya and the Netherlands, to whom I am greatly indebted. During the process of conducting this research, I had the pleasure of becoming friends with some of these people; friendships that I intend to keep and cherish beyond my PhD studies.

Special thanks to my fellow PhDs Abby and Alex, with whom I had many memorable moments. Discussions and occasional sharing of frustrations with you helped me feel I am not alone and kept me going. Alex, I am also extremely grateful to you for sharing your local knowledge and insights and for helping me with arranging interviews during both data collection phases of my research in Kenya.

I would also like to particularly thank my colleague and my dear friend Onno for translating the summary of this dissertation to Dutch.

Friends at IHE and beyond; I know some of you from a long time ago and I was lucky enough to get to know some of you during my PhD studies. You have been a major source of support for me during this journey and I have wonderful memories from spending time with each and every one of you. You are all dear to me, so I will keep the list alphabetical: Abdi, Afnan, Ahmad, Akosua, Ali D., Ali N., Andres, Ane, Azadeh, Babak, Behnood, Berend, Cath, Chris, Doro, Ehsan, Elisa, Eric, Eva, Farhad, Fernanda, Flora, Hadi, Juan Carlos, Juliette, Linda, Luana, Mahshid, Masoud, Mauri, Megan, Meike, Mina, Mohammad Reza, Mohan, Mohaned, Mojtaba P., Mojtaba S. Navid, Pantea, Poolad, Rassoul, Roxana, Sachin, Sanaz, Sara, Somayeh, Sonia, Thaine, Vahid and Yared.

I cannot begin to express my gratitude to my mother Maryam and my late father Abdolali whose devotion, unconditional love and support has made me the person that I am today. I am extremely grateful to you for going above and beyond with what you could possibly do for your children. I also dearly appreciate all the support and encouragement that I received from my siblings Behnam, Nazi, Behrouz and Bahman along this way. I also would like to thank my mother in-law Fereshteh, my late father in-law Kazem and their children Mohammad and Sara who have encouraged Bahar and me in every aspect of our lives.

The last words are reserved for Bahar, the love of my life and my best friend. This journey was by no means possible without your encouragements, selflessness, sacrifices and never-ending support. In order to make this a success, you had to quit your job, move to a different country, learn a new language and experience a lot of ups and downs along the way. Regardless of the circumstances, you always kept your smile, gave me hope and helped me to keep my sanity through tough times. You are a very special person and have an incredibly strong character. I am forever indebted to you and I know together we can achieve anything we set our minds to.

SUMMARY

Citizen participation in water and environmental management via community-based monitoring (CBM) initiatives has been praised for the potential to facilitate better informed, more inclusive, transparent, and representative decision making. However, the conceptual understanding required to critically analyze and understand the dynamic processes that might lead to such promised effects and the short, medium and long term outputs and impacts of these processes, is largely limited. This is due to the fact that there have not (yet) been enough instances of methodological and empirical research that try to conceptualize and evaluate these dynamics and outcomes of CBM initiatives.

The main objective of this research was to conduct a systematic evaluation of the factors that influence the establishment, functioning and outcomes of CBM initiatives. This was done using a qualitative empirical research methodology and by employing a case study approach. This research was conducted in the context of the Ground Truth 2.0 project and therefore two of the six Demo Cases of this project were selected as the case studies of this research; namely, case studies in Kenya and the Netherlands. The CBM initiative developed in the Netherlands is called Grip op Water Altena and focuses on the issue of pluvial floods in 'Land van Heusden en Altena'. The Kenyan CBM is called Maasai Mara Citizen Observatory and aims at contributing to a better balance between biodiversity conservation and sustainable livelihood management in the Mara ecosystem.

In line with the main objective of the study, and based on the review of a large body of literature in the fields of community-based monitoring, Citizen Science and affiliated fields of research, combined with the empirical evidence from a number of past EU-funded CBM projects, a framework was developed that guided the empirical evaluations of this research. The distinction between five different dimensions, and 22 internal and context-related factors, is a unique feature of this framework that broadens its applicability and makes it suitable for 'Context analysis', 'Process evaluation' and 'Impact assessment' of CBM initiatives. The introduced framework is therefore called the CPI Framework in short. Studying a CBM using the CPI Framework provides an interpretation of what 'community' means in the context of a CBM initiative; a concept that is difficult to depict and study otherwise.

The empirical evaluation of the establishment dynamics and results of the case studies of this research was conducted using a two phase design approach.

In the first phase, the CPI Framework was used for conducting a systematic analysis of the baseline situation of two case studies, before establishment of the two CBM initiatives. The aim of this phase of the research was to gain a thorough understanding of the social, institutional, political and technological contexts in which these CBMs were going to be established and with which they would interact. This baseline analysis showed that aside from the fact that these initiatives have different thematic foci, there are distinct differences in terms of access to technology, availability and accessibility of data, the institutional arrangements for public participation in decision making processes, and the level of citizen trust in the authorities in charge of managing the respective water-related and environmental issues.

The second phase of this research focused on using the CPI Framework for conducting a systematic evaluation of the establishment process and results of Grip op Water Altena and Maasai Mara Citizen Observatory. This allowed for both a detailed analysis of each CBM as well as a cross-case analysis of the factors that affected the establishment and functioning of the two CBMs.

The findings of this study demonstrate that systematic evaluation of goals and objectives, participation processes, power dynamics, technological choices and results of CBMs can indeed provide critical insights into the establishment process and functioning of these initiatives. Moreover, factors influencing the establishment process and functioning of CBMs are not only internal to the initiatives, but also context-related. Conducting a baseline analysis before or at early stages of establishment of a CBM can help enhance understanding of the contextual realities in which the CBM will operate and provides a basis for measuring its outcomes and impacts

The results of a two phase empirical research into the factors that influenced the establishment and functioning of Grip op Water Altena and Maasai Mara Citizen Observatory shows that CBMs should strive for realistic and specific objectives and carefully consider actor-specific interests and contextual settings that may enable or hinder achieving those objectives. This is especially important in the case of CBMs that aim at moving beyond the environmental monitoring function and engage with policy and decision making processes. Solving complex environmental challenges or balancing existing and un-even power relationships between stakeholders is far from easy. CBMs should therefore be power-sensitive in their process of establishment and realistically assess if and to what extent they can contribute to solving such complex problems.

Perceived urgency or importance of the topic, existing power relationships, level of trust among the actors, length of the establishment process and ease or difficulty of participation are factors that affect the initial and continued participation of stakeholders in a CBM. Moreover, establishing CBMs in developing countries and regions with limited technological advancements is particularly challenging and requires careful considerations for inclusion of vulnerable and less tech-savvy community members. Compatibility of technological choices with social, institutional and technological context reduces the chance of excluding major groups within society. Nevertheless, heterogeneity of society should be acknowledged and realistic expectations should be set and communicated about the extent to which CBMs can enable participation of different groups within society. The study also demonstrated that data, information and knowledge exchange, awareness raising, learning opportunities, and communication and interaction possibilities created because of a CBM are among the more immediate, tangible and easier to study results of CBMs. In contrast, environmental impacts and shift in power-relationships among stakeholders are more long-term. Given the design of this research, these longer term outcomes and impacts could not be studied.

The initial need for establishing funded project-driven CBMs does not usually come from the local stakeholders and most often the idea for establishing these initiatives comes from researchers and funders. This increases the chance that these CBMs are more 'supply-driven' than 'demand-driven'. Moreover, the establishment process of project-driven CBMs is very likely to be influenced by factors such as pre-framing of the issue and scope of the initiative, as well as pre-defined resources, time-frame and other obligations towards funding organizations.

Study of the factors that influenced the establishment and functioning of Grip op Water Altena and Maasai Mara Citizen Observatory also generated insights that are especially important for co-designed CBMs. For example, establishing CBMs using a co-design approach is a time-demanding and resource-intensive process that requires efforts and commitment from all involved actors. CBMs that follow a co-design methodology should set a clear timeframe for defining their aims, objectives and functionalities and participants in the co-design process should be made aware of the time commitment they need to make for participation. Moreover, a co-design process provides possibilities for discussion and consensus building among different stakeholders and thus provides a more equal chance for parties involved to influence the establishment processes of a CBM. Nevertheless, the fact that a CBM is co-created or co-designed does not mean that power relationships between stakeholders do not exist or are balanced out completely.

In summary, this dissertation contributes to enhancing both conceptual and empirical understanding of CBMs in a number of ways. First, it contributes to conceptualization of CBMs by developing the CPI Framework that is suitable for context analysis, process evaluation and impact assessment of CBM initiatives. This conceptualization is built on theoretical and empirical evidence from literature and lessons learned from the establishment of CBMs in the context of five 'pioneer' or 'legacy' EU-funded projects. Second, a major contribution of this dissertation to empirical understanding of CBMs is a detailed picture of the establishment process and the results of two real life project-based CBMs; one in Europe and one in Africa. This detailed picture built on perspectives of both local stakeholders who participated in the establishment of Grip op Water Altena and Maasai Mara Citizen Observatory, as well as members of the Ground Truth 2.0 team who were involved in establishing the two CBMs, and therefore allowed for comparing and contrasting the perceptions of these two distinct groups.

SAMENVATTING

Burger participatie in water- en milieubeheer via community-based monitoring (CBM) initiatieven is geprezen voor het potentieel om beter geïnformeerde, meer inclusieve, transparante en representatieve besluitvorming mogelijk te maken. Het conceptuele inzicht dat vereist is om de dynamische processen die kunnen leiden tot dergelijke beloofde effecten, en de korte, middellange en lange termijn uitkomsten en impacts van deze processen, kritisch te kunnen analyseren en te begrijpen, is echter grotendeels beperkt. Dit komt door het feit dat er (nog) onvoldoende voorbeelden van methodologisch en empirisch onderzoek is geweest die deze dynamiek en resultaten van CBM-initiatieven proberen te conceptualiseren en evalueren.

Het hoofddoel van dit onderzoek was om een systematische evaluatie uit te voeren van de factoren die van invloed zijn op de oprichting, werking en resultaten van CBM-initiatieven. Dit werd gedaan met behulp van een kwalitatieve empirische onderzoeksmethode en door een casus-aanpak te gebruiken. Dit onderzoek werd uitgevoerd in het kader van het Ground Truth 2.0-project en daarom werden twee van de zes demo casussen van dit project geselecteerd als casussen van dit onderzoek; namelijk de casussen in Kenia en Nederland. Het CBM-initiatief dat in Nederland is ontwikkeld, heet Grip op Water Altena en richt zich op de kwestie van wateroverlast in 'Land van Heusden en Altena'. Het Keniaanse CBM heet Maasai Mara Citizen Observatory en heeft als doel bij te dragen aan een beter evenwicht tussen behoud van biodiversiteit en duurzaam beheer van levensonderhoud in het Mara-ecosysteem.

In overeenstemming met de hoofddoelstelling van de studie, en op basis van de beoordeling van een grote hoeveelheid literatuur op het gebied van community-based monitoring, Citizen Science en aanverwante onderzoeksgebieden, gecombineerd met empirisch bewijs uit in het verleden gefinancierde CBM-projecten in een aantal EU-landen, werd een raamwerk ontwikkeld dat de empirische evaluaties van dit onderzoek leidde. Het onderscheid tussen vijf verschillende dimensies en 22 interne en context gerelateerde factoren is een uniek kenmerk van dit raamwerk dat de toepasbaarheid ervan verbreedt en het geschikt maakt voor 'Contextanalyse', 'Procesevaluatie' en 'Impactanalyse' van CBM-initiatieven. Het geïntroduceerde raamwerk wordt daarom in het kort het CPI-raamwerk genoemd. Het bestuderen van een CBM met behulp van het CPI Framework geeft een interpretatie van wat 'gemeenschap' betekent in de context van een CBM-initiatief; een concept dat anders moeilijk is weer te geven en te bestuderen.

De empirische evaluatie van de vestigingsdynamiek en de resultaten van de casussen van dit onderzoek werd uitgevoerd met behulp van een benadering die bestaat uit twee fasen.

In de eerste fase werd het CPI-kader gebruikt om een systematische analyse van de basissituatie van de twee casussen uit te voeren, voordat de twee CBM-initiatieven werden opgezet. Het doel van deze fase van het onderzoek was om een grondig inzicht te krijgen in de sociale, institutionele, politieke en technologische context waarin deze CBM-initiatieven zouden opgericht worden, en waarmee ze zouden interageren. Deze basisanalyse toonde aan dat er, afgezien van het feit dat deze initiatieven verschillende thematische aandachtspunten hebben, er duidelijke verschillen zijn wat betreft toegang tot technologie, beschikbaarheid en toegankelijkheid van gegevens, de institutionele regelingen voor publieke participatie in besluitvormingsprocessen en het niveau van de burgerlijke vertrouwen in de autoriteiten die verantwoordelijk zijn voor het beheer van de water gerelateerde- en milieukwesties in de twee casussen.

De tweede fase van dit onderzoek was gericht op het gebruik van het CPI-kader voor het uitvoeren van een systematische evaluatie van het vestigingsproces en de resultaten van Grip op Water Altena en Maasai Mara Citizen Observatory. Dit maakte zowel een gedetailleerde analyse van elke CBM mogelijk, als een analysen tussen de casussen van de factoren die van invloed waren op de oprichting en werking van de twee CBM's.

De bevindingen van deze studie tonen aan dat systematische evaluatie van doelen en doelstellingen, participatieprocessen, machtsdynamiek, technologische keuzes en resultaten van CBM's inderdaad kritische inzichten kunnen verschaffen in het vestigingsproces en de werking van deze initiatieven. Bovendien zijn factoren die het vestigingsproces en het functioneren van CBM's beïnvloeden niet alleen intern in de initiatieven zijn, maar ook context gebonden zijn. Het uitvoeren van een nulmeting vóór, of in de vroege stadia van de oprichting van een CBM, kan helpen met het begrijpen van de contextuele realiteit waarin de CBM zal werken en biedt een basis voor het meten van de resultaten en effecten ervan.

De resultaten van een tweefasig empirisch onderzoek naar de factoren die de oprichting en werking van Grip op Water Altena en Maasai Mara Citizen Observatory hebben beïnvloed, tonen aan dat CBM's moeten streven naar realistische en specifieke doelstellingen, en zorgvuldig rekening houden met actor specifieke belangen en contextuele factoren die mogelijk het bereiken van die doelstellingen kunnen belemmeren. Dit is vooral belangrijk in het geval van CBM's die erop gericht zijn verder te gaan dan de milieumonitoringfunctie en zich bezig houden met beleids- en besluitvormingsprocessen. Het oplossen van complexe milieu-uitdagingen, of het in evenwicht brengen van bestaande en onevenwichtige machtsverhoudingen tussen belanghebbenden, is verre van eenvoudig. CBM's moeten daarom vermogensgevoelig zijn in hun vestigingsproces en realistisch beoordelen of, en in welke mate, ze kunnen bijdragen aan het oplossen van dergelijke complexe problemen.

Waargenomen urgentie of importantie van het onderwerp, bestaande machtsverhoudingen, niveau van vertrouwen tussen de actoren, duur van het vestigingsproces en gemak of moeilijkheid van deelname, zijn factoren die de initiële en voortdurende deelname van belanghebbenden aan een CBM beïnvloeden. Het opzetten van CBM's in ontwikkelingslanden en regio's met beperkte technologische vooruitgang is bovendien bijzonder uitdagend en vereist zorgvuldige overwegingen voor de opname van kwetsbare en minder technisch begaafde leden van de gemeenschap. Compatibiliteit van technologische keuzes met sociale, institutionele en technologische context verkleint de kans om grote groepen in de samenleving uit te sluiten. Niettemin moet de heterogeniteit van de samenleving worden erkend en moeten realistische verwachtingen worden gesteld en gecommuniceerd over de mate waarin CBM's deelname van verschillende groepen in de samenleving mogelijk maken. De studie toonde ook aan dat gegevens, informatie en kennisuitwisseling, bewustmaking, leermogelijkheden en communicatie- en interactiemogelijkheden die door een CBM gecreëerd worden, behoren tot de meer directe, tastbare en gemakkelijker te bestuderen resultaten van CBM's. Milieueffecten en verschuivingen in machtsverhoudingen tussen belanghebbenden zijn daarentegen van langere duur. Gezien de opzet van dit onderzoek konden deze uitkomsten en effecten op langere termijn niet worden bestudeerd.

De initiële behoefte aan het opzetten van gefinancierde project gestuurde CBM's komt meestal niet van de lokale belanghebbenden. Meestal komt het idee voor het opzetten van deze initiatieven van onderzoekers en financiers. Dit vergroot de kans dat deze CBM's meer 'aanbodgestuurd' zijn dan 'vraaggestuurd'. Bovendien wordt het vestigingsproces van project gestuurde CBM's zeer waarschijnlijk beïnvloed door factoren zoals het vooraf bepalen van het probleem en de reikwijdte van het initiatief, evenals vooraf gedefinieerde middelen, tijdsbestek en andere verplichtingen tegenover financieringsorganisaties.

Onderzoek naar de factoren die de oprichting en werking van Grip op Water Altena en Maasai Mara Citizen Observatory hebben beïnvloed, heeft ook inzichten opgeleverd die vooral belangrijk zijn voor mede-ontworpen CBM's. Het opzetten van CBM's met behulp van een co-designbenadering is bijvoorbeeld een tijdrovend en middel intensieve proces dat inspanningen en inzet van alle betrokken actoren vereist. CBM's die een co-designmethodiek volgen, moeten een duidelijk tijdschema vaststellen voor het definiëren van hun doelen, doelstellingen en functionaliteiten, en deelnemers aan het co-designproces moeten bewust worden gemaakt van de tijdsbesteding die ze moeten doen voordat ze deelnemen. Bovendien biedt een co-ontwerpproces mogelijkheden voor discussie en consensusvorming tussen verschillende stakeholders en biedt het dus een meer gelijke kans voor betrokken partijen om de vestigingsprocessen van een CBM te beïnvloeden. Het feit dat een CBM mede is gecreëerd of mede is ontworpen, betekent echter niet dat machtsverhoudingen tussen belanghebbenden niet bestaan of volledig in evenwicht zijn.

Dit proefschrift draagt op een aantal manieren bij aan het verbeteren van zowel conceptueel als het empirisch begrip van CBM's. Ten eerste draagt het bij aan de conceptualisering van CBM's door het CPI Framework te ontwikkelen dat geschikt is voor contextanalyse, procesevaluatie en effectbeoordeling van CBM-initiatieven. Deze conceptvorming is gebaseerd op theoretisch en empirisch bewijs uit de literatuur, en op basis van de lessen die zijn getrokken uit de oprichting van CBM's in de context van vijf door de EU gefinancierde projecten die 'pionier' of 'legaat' zijn. Ten tweede is een belangrijke bijdrage van dit proefschrift aan empirisch begrip van CBM's door een gedetailleerd beeld van het vestigingsproces te schetsen door de resultaten van twee levensechte project gebaseerde CBM's; één in Europa en één in Afrika. Dit gedetailleerde beeld bouwde voort op de perspectieven van de lokale belanghebbenden die hebben deelgenomen aan de oprichting van Grip op Water Altena en Maasai Mara Citizen Observatory, evenals leden van het Ground Truth 2.0-team die betrokken waren bij de oprichting van de twee CBM's. Hierdoor konden de percepties van deze twee verschillende groepen vergeleken en gecontrasteerd worden.

CONTENTS

1

INTRODUCTION[1]

[1] This chapter is partially based on the following publications:

Gharesifard, M., Wehn, U., & van der Zaag, P. (2019b). What influences the establishment and functioning of community-based monitoring initiatives of water and environment? A conceptual framework. *Journal of Hydrology, 579*, 124033. doi:https://doi.org/10.1016/j.jhydrol.2019.124033

Gharesifard, M., Wehn, U., & van der Zaag, P. (2017). Towards benchmarking citizen observatories: Features and functioning of online amateur weather networks. *Journal of Environmental Management, 193*, 381-393. doi:https://doi.org/10.1016/j.jenvman.2017.02.003

1.1 BACKGROUND

This section provides background information relevant for the topic of this doctoral research. It starts with a short introduction about the environmental challenges in the 21st century and the wider topic of citizen participation in environmental management and governance. Although stakeholder participation is generally accepted as a good practice in environmental management and governance, its purpose, added value, necessity, and the process that leads to it has raised a number of questions. Therefore, the next section is dedicated to discuss the paradoxes about citizen participation in environmental management. The subsequent two sections introduce the concepts of citizen science and community-based monitoring initiatives that is the core focus of this research.

1.1.1 Environmental management in the 21st century

There is an ever-increasing competition over limited natural resources. According to the United Nation's most recent report on global population prospect; by 2030, we will share the limited natural resources of our planet with approximately 800 million more people (United Nations, 2019). Next to population growth, a number of environmental challenges increasingly affect our ecosystem and place fundamental threats on human well-being and quality of life and also undermine peace and development (United Nations, 2012). The most recent World Economic Forum's yearly report on global landscape of risks identifies environmental challenges such as 'extreme weather events', 'natural disasters' and 'failure of climate-change mitigation and adaptation' among the top 5 global risks both in terms of likelihood of occurrence and global impact (World Economic Forum, 2019).

Facing these global challenges and moving towards a sustainable future, requires improved policies and informed environmental decision making. On the one hand, a pre-requisite for better informed environmental decision making is continuous and widespread observations of the environment that can generate required data to inform policies. OECD's Environmental Outlook 2050 suggests that "better information supports better policies, so our knowledge base needs to be improved" (OECD, 2012, p. 8). One may argue that availability of more data and information does not guarantee better environmental policies. Nonetheless, absence or limited availability of data and information often results in increased uncertainties in decision making processes. On the other hand, there is an ever increasing recognition that environmental science and policy should be more participatory, transparent and democratic. It means that the environmental sciences should be opened up to the public and incorporate locally-relevant knowledge. In addition, environmental decisions should consider the voice of citizens whose well-being and livelihood are being affected by those decisions.

1.1.2 Citizen participation in environmental management and governance

During the past two decades, there has been an increasing recognition that structural environmental measures (e.g. building dams or flood defense infrastructure) cannot be the solution for complex environmental challenges of the 21st century. Therefore, non-structural environmental management measures such as development and upgrading of flood early warning systems, improved land-use planning, flood proofing, insurance and awareness campaigns have been highly promoted as hybrid approaches to environmental management problems (Bradford et al., 2012; Wehn et al., 2015b; Yamada et al., 2011).

The promotion of non-structural measures along with a widespread replacement of environmental management concepts with governance ideology (Wehn et al., 2015b) has highlighted the role of citizens as one of the most important stakeholders. This has resulted in a particular attention to engaging citizens with environmental management practices. It has been argued that "if organized well, public participation can result in valuable information for planners and decision-makers" (Mokorosi & van der Zaag, 2007, p. 324). As a result, the importance of citizen engagement with environmental management has been recognized and highlighted in various international policy guidelines. Box 1 provides the link to a number of these policy guidelines.

BOX 1: Citizen engagement with environmental management in policy guidelines

- Principle 10 of the Rio declaration, states that *"Environmental issues are best handled with participation of all concerned citizens, at the relevant level. At the national level, each individual shall have appropriate access to information concerning the environment that is held by public authorities, including information on hazardous materials and activities in their communities, and the opportunity to participate in decision-making processes. States shall facilitate and encourage public awareness and participation by making information widely available. Effective access to judicial and administrative proceedings, including redress and remedy, shall be provided"* (UNDP, 1992).

- All parties of the 1998 Aarhus (UNECE) convention recognized that, *"in the field of the environment, improved access to information and public participation in decision-making enhance the quality and the implementation of decisions, contribute to public awareness of environmental issues, give the public the opportunity to express its concerns and enable public authorities to take due account of such concerns"* (UNECE, 1998).

- The Hyogo Framework for Action (HFA) in its 3rd Priority Action emphasizes that *"Disasters can be substantially reduced if people are well informed and motivated towards a culture of disaster prevention and resilience, which in turn requires the collection, compilation and dissemination of relevant knowledge and information on hazards, vulnerabilities and capacities"* (UNISDR, 2005).

- Different targets of the SDGs have particular reference to stakeholder participation, and especially participation of vulnerable and excluded groups (including target 5.5, 6.b and 10.6), however, target 16.7 explicitly aims to *"ensure responsive, inclusive, participatory and representative decision-making at all levels"* (United Nations, 2015).

1.1.3 Paradoxes of citizen participation in practice

Different instances of scientific research and policy documents have multiple and sometimes contradictory views about the purpose of citizen engagement, its necessity and added value in different contexts, the engagement process itself, and the outcomes and impacts of such processes. This has given rise to critical questions about participatory approaches and forms a number of paradoxes that are discussed in this section.

Three different goals have been identified and discussed in the literature for promoting citizen engagement (Cleaver, 1999; Kruger, 2010; Mayoux, 1995; Nelson & Wright, 1995; World Economic Forum, 2016b); (1) The *'efficiency'* goal that looks at citizen engagement as a tool for improved public services and project outcomes, (2) the *'empowerment and equity'* purpose that is closely linked to the debates about democratization and providing the general public with opportunities to learn and express their voice in decisions, and (3) the *'social stability'* purpose that is built on the belief that citizen engagement will help reduce conflicts and social unrest.

Cleaver (1999) criticizes the way efficiency and empowerment arguments are made and mentions that these concepts are often defined and presented in a depoliticized way and thus do not help answer critical questions such as improved efficiency for whom? Who is targeted to be empowered? And how does the process affect certain individuals or groups within society (e.g. women, low-income citizens, elderly, and socially excluded people)? The same argument could be made for the social stability purpose: how can a participatory process ensure that it does not exclude certain groups within society and hence create new forms or even increase social unrest? Thus, it is essential to identify social and political power relations and understand how decisions are being taken in the context of each participatory setting in order to avoid creating hollow perceptions about the added value of public engagement. In this regard, Warner (2006) conducted a study on multi-stakeholder platforms for integrated catchment management using several case studies (Peru, Argentina, India, South Africa and Belgium) and concluded that none of these platforms had a significant mandate and no real power sharing took place as a result of these participatory processes.

Several studies claim that the information flow that takes place in a participatory process will result in more transparent decisions and hold government accountable (Bertot et al., 2010; Grandvoinnet et al., 2015; Nabatchi, 2012; Reed, 2008; Videira et al., 2006; Warburton et al., 2001; World Bank Group, 2014). This is especially more emphasized in interactive settings where citizens and decision makers communicate and share information and requests. However, claims that this increased inclusiveness, transparency and accountability will result in improved living conditions for citizens (especially the poor and marginalized citizens), has been criticized by some scholars (Cleaver, 1999; Davenport, 2013; Guijt, 2014; Warner, 2006). Warner explains that "for some stakeholders, the communication and information process itself is good enough, but others will want results: 'food on the table'" (Warner, 2006, p. 15).

Another paradox of citizen engagement is linked to the conceptualization of 'the community'. In this regard, participatory projects have been criticized for imagining improved outcomes for society as a whole; communities that are heterogeneous by nature and include people with varying motivations, beliefs, and livelihoods, who will be affected differently (Cleaver, 1999; Warner, 2006). In a study that focused on the gender aspects of public engagement, Mayoux (1995) concluded that different stakeholders and individuals have different agendas and thus chances for reaching a consensus that does not affect anyone are very low. The result of this heterogeneity often affects the weaker strata in society and excludes them from the decision making processes (Flyvbjerg, 1998;

5

McGuirk, 2001). Although it is not possible to have an in-depth understanding of how every single individual in society might be affected in a participatory setting, it is critical to study and understand how different groups are affected and how a specific participatory setting aligns/collides with their motivations, beliefs, or livelihoods.

1.1.4 Citizen science: what it is and why it matters

The term 'citizen science' has been used to describe a spectrum of participatory processes with the aim of studying natural phenomena that often involves collaboration between citizen, scientists and (less frequently) decision makers.

The first incident of using the term 'citizen science' was recorded in January 1989 when 225 volunteer citizens from all states of United States of America took part in a 'citizen science' program which involved collecting rain samples, testing their acidity and reporting the results to Audubon headquarters that was in charge of publishing the national map of acid-rain levels (Haklay, 2014). However, 'citizen science' is much older than the creation of its name tag and its initiation dates back to the 18th century (McCarthy et al., 2013). There have been a number of attempts to classify citizen science initiatives. Among others, Wiggins and Crowston (2012) classified these initiatives based on the stated project goals and tasks performed by participants; Haklay (2015) proposed a classification based on the level of engagement and commitment of participant; and Kullenberg and Kasperowski (2016) in a meta-analysis of citizen science literature clustered these initiatives based on their higher purpose. The following three categories are based on Kullenberg and Kasperowski (2016) and help introducing different forms of citizen science.

(1) Citizen science as a method

This category is perhaps the most common form of citizen science that has recently gained significant momentum in natural resources studies (Kullenberg & Kasperowski, 2016). The definition for this category is well-captured by the Oxford English Dictionary; "the collection and analysis of data relating to the natural world by members of the general public, typically as part of a collaborative project with professional scientists" (Oxford English Dictionary, 2014). Another definition for this category of citizen science initiatives is proposed by OpenScientist blog, where they define 'citizen science' as "the systematic collection and analysis of data; development of technology; testing of natural phenomena; and the dissemination of these activities by researchers on a primarily vocational basis" (OpenScientist blog, 2011). These definitions advocate citizen science as a method for gathering, classifying and analyzing data, which will be further processed and used by the scientists and policy makers.

(2) Citizen science as public engagement with science and policy

This is a notion that originates from social and political science and was influenced by Alan Irwin's book 'Citizen Science: A Study of People, Expertise and Sustainable Development' from 1995 (Kullenberg & Kasperowski, 2016). Irwin depicted citizen science as "a science which assists the needs and concerns of citizens ... [and] at the same time [is] a form of science developed and enacted by citizens themselves" (Irwin, 1995, p. xi). This definition moves beyond the passive form of citizen science explained in the first form and perceive a more active role for citizens that can potentially influence the decision making processes in a more practical way (Bonney et al., 2015; Dickinson et al., 2012; Kullenberg & Kasperowski, 2016; Wehn & Evers, 2014).

(3) Citizen science as civic mobilization

The third form of citizen science is initiated by citizens themselves and is often triggered by issues of concern for their communities. These issues often relate to environmental issues (e.g. water and air pollution, species conservation, health hazards), but also may have the purpose of opposing decisions made by authorities. The higher aim of these social movements is to gain legal or political influence in matters of concern through joint action, evidence gathering, and awareness-raising.

Regardless of the classifications used for defining/understanding citizen science initiatives, sometimes it is very difficult to draw a clear cut line and place citizen science projects into these distinct categories. In other words, in reality these lines are blurry and one can find various grey areas while trying to categorize citizen science projects.

Research on 'citizen science' boomed during the last decade. Silvertown (2009) stated that 80 % of the articles related to 'citizen science' existing in the Web of Knowledge were published between 2005 and 2009. Anne Bowser and Elizabeth Tyson from the Commons Lab, Wilson Center (in the foreword section of Haklay (2015)) claim that one reason behind the recent attractiveness of citizen science projects is the expected production of large-scale and cost-effective data in these initiative. Because of this efficiency-related added value (section 1.1.3 above), citizen science projects are expected to result in budgetary cuts and help bolster limited and declining, governmental and organizational resources (Haklay, 2015). This statement may be partially true, but only reflects the benefits of 'passive citizen science' (Nature, 2015; Wehn & Evers, 2014) projects or the first form of citizen science as introduced above (i.e. citizen science as method). However, if we perceive a more 'active role' for citizens in governance and decision making processes (Nature, 2015; Wehn & Evers, 2014), citizen engagement might have the potential to promote inclusive, transparent and accountable decision making in different domains (Bonney et al., 2015; Dickinson et al., 2012; Gigler & Bailur, 2014; Wehn et al., 2015b) and even help promote social stability in societies (World Economic Forum, 2016b).

1.1.5 Community-based monitoring initiatives

A number of researchers have categorized citizen science projects into different typologies (Cortes Arevalo, 2016; Ferster & Coops, 2013; Haklay, 2013). One of these typologies is community-based monitoring initiatives of the environment (hereafter CBM). CBM is "a process where concerned citizens, government agencies, industry, academia, community groups and local institutions collaborate to monitor, track and respond to issues of common community concern" (Whitelaw et al., 2003, p. 410). As emphasized by Conrad and Hilchey (2011), CBM refers to both community-based environmental monitoring and community-based environmental management aspects of Citizen Science. This definition is very close to what is also referred to as 'citizen observatories of the environment'. The concept of 'citizen observatory' is mostly used in the European context and the European Commission defines it as "community-based environmental monitoring and information systems using innovative and novel earth observation applications" such as portable devices (e.g. smart phones) and collective intelligence to support both community and policy priorities (European Commission, 2014; Lanfranchi et al., 2014; Rubio Iglesias, 2015). CBMs have been praised for their potential to contribute to better environmental decision-making by empowering citizens and allowing them to take a more active role in environmental monitoring, co-operative planning and environmental stewardship (European Commission, 2015).

As the definition of citizen observatories clarifies, Information Communication Technologies (ICTs) play a key role in these initiatives. Recent technological developments and advancements in ICTs have transformed citizen science and CBMs to a great extent as they have provided stakeholders with new possibilities for data collection, sharing and communication.

CBM initiatives have been conceptualized as interactive settings, in which citizens, data aggregators/scientists, and policy/decision makers communicate with each other (Wehn et al., 2015a). Figure 1.1 depicts this conceptualization; it illustrates the interaction between different stakeholder groups and the distinct ways by which citizens can possibly play a role in environmental decision making that ranges from (implicit and explicit) data collection to cooperative planning and environmental stewardship. The timeline below Figure 1.1 indicates that closing the loop and creating an interactive dialogue between citizens and decision makers is a time-based process. It requires two-way communication and close collaboration between citizens, data aggregators and decision makers.

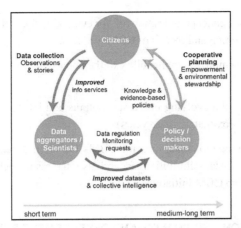

Figure 1.1 Interrelations between different stakeholders in a CBM initiative
Source: Wehn et al. (2015a)

CBMs can present a potential way forward in enhancing decision making processes and facing prominent environmental challenges of the 21st century. However, due to the novelty of the concept, there is still a need for studying and better understanding if and how this potential added value of CBM can be translated into real enhancements of environmental management practices.

1.2 PROBLEM STATEMENT AND RESEARCH OBJECTIVES

CBM initiatives have been praised for their potential to facilitate better informed, more inclusive, transparent, and representative environmental decision making. Despite this potential, successful establishment of these initiatives has proven to be an intricate task. For example, initial and long term engagement of different stakeholders in CBM activities is still a huge challenge and the actual added value of CBM for enhanced environmental decision making has remained a controversial issue. Practitioners and scientists in the field of citizen science have been mostly preoccupied with promoting CBM initiatives as a tool for producing more environmental data. This has resulted in an evident gap in the quantity and quality of analytical research that tries to study CBM initiatives and the factors that influence their establishment, functioning and outcomes. It is therefore essential to take the time to learn from the past experiences and critically document, analyze and understand internal and context-related factors that might influence the establishment, functioning and outcomes of CBM initiatives.

The main objective of this research is therefore to conduct a systematic evaluation of the factors that influence the establishment and functioning of CBM initiatives, as well as direct products and changes resulting from their establishment. In line with the main objective of the study, the specific research objectives are as follows:

9

Objective 1: Develop a conceptual framework for evaluating the factors that influence the establishment, functioning and results of CBM initiatives.

Objective 2: Test the empirical applicability of the conceptual framework by conducting a baseline analysis of two CBM initiatives.

Objective 3: Evaluate the evolving processes, outputs and interim outcomes of the two CBMs over time (approximately three years).

Objective 4: Provide recommendations for CBMs based on a detailed analysis of the characteristics of (un)successful initiatives, and also, the results achieved and obstacles experienced by the two CBM initiatives.

1.3 CONTRIBUTION, ORIGINALITY AND DEVELOPMENT RELEVANCE

This research builds on a large body of theoretical and empirical studies in the fields of Citizen Science, Science and Technology Studies (STS), public participation in decision making processes and e-participation, i.e. ICT-enabled participation in governance processes (Jafarkarimi et al., 2014; Macintosh & Coleman, 2003). It synthesizes this cross-disciplinary knowledge into a much needed conceptual framework which can serve to unpack different factors that influence the establishment and functioning of CBMs. Furthermore, the framework developed in this research is applied to analyze the factors that affect the establishment and functioning of two CBM initiatives in the Netherlands and Kenya. This allowed for a cross-case analysis of the factors that affected the establishment and functioning of two CBMs in the context of a 'developed' versus a 'developing' country.

The objectives and focus of this research are closely related to the globally adopted 2030 agenda for sustainable development (Sustainable Development Goals). The role of citizen science in supporting authorities to fill data gaps needed to achieve the SDGs is now recognized more than ever (Fritz et al., 2019; Lu et al., 2015). To do so, societies need to transform and citizens need to take a more active role in monitoring their living environment. Recent advancements in ICTs have enabled a far deeper and faster process of such transformation in society as compared to the past (Sachs et al., 2015). Moreover, ICT-enabled community-based monitoring can provide an enabling environment for citizens to interact with each other, the private sector, and government and this interaction might open avenues for citizens to play a crucial role in development-related decision-making processes (Gigler & Bailur, 2014; Hsu et al., 2014; Wehn et al., 2015b). In the area of natural resources management, these avenues include participation in data collection, cooperative planning and environmental stewardship (Wehn et al., 2015a). Evaluating the dynamic processes that influence the establishment, functioning and outcomes of CBM initiatives provides opportunities to better understand whether (and to what extent) citizen engagement promotes "responsive, inclusive, participatory and representative decision-making" (as emphasized in the target 16.7 of the SDGs), and, more importantly, how to facilitate such positive outcomes.

1.4 OUTLINE OF THE DISSERTATION

This dissertation is structured as follows. Chapter 2 provides the theoretical context of this research based on an extensive review of relevant literature in the fields of Citizen Science, e-participation, STS and public participation in decision making processes. Furthermore, Chapter 2 introduces the conceptual framework of the research that was developed based on the results of the aforementioned literature review, and presents the research questions of this study. The two phase methodology that was followed for data collection and analysis of this research is detailed in Chapter 3. Chapters 4 and 5 are dedicated to presenting the findings of the empirical research that was conducted in the two case studies of this research in the Netherlands and Kenya. Chapter 6 presents the results of a cross-case analysis that was performed to compare and contrast the most important factors that influenced the establishment, functioning and outputs of the two CBMs. Finally, Chapter 7 provides the conclusions of the research. This chapter includes a reflection on the methodological and theoretical choices that were made for conducting this research and provides a series of recommendation for policy makers, scientists and citizens, based on the findings of this study.

Figure 1.2 presents a summary of the outline of the dissertation along with the linkage between the chapters.

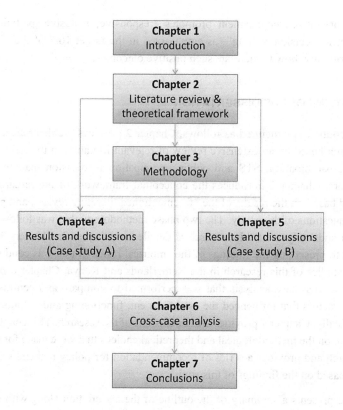

Figure 1.2 Outline of the dissertation

2

LITERATURE REVIEW AND CONCEPTUAL FRAMEWORK[2]

Chapter 1 provided the introduction about the topic of this dissertation and introduced the objectives of this research. In line with the first objective (i.e. developing a conceptual framework for evaluating CBMs), this chapter presents the conceptual framework of this research. This framework builds on a large body of literature in the fields of citizen science and other affiliated fields of research. Therefore, section 2.1 is dedicated to presenting the literature review that was conducted for developing the conceptual framework of this doctoral dissertation and an introduction to the emergent dimensions from this review. Section 2.2 elaborates on the identified dimensions and provides the link between the in-depth discussions about each dimension and the reviewed literature. A summary of the conceptual framework is provided in section 2.3. Based on the objectives of the research that were introduced in Chapter 1 and the introduced conceptual framework in this chapter, section 2.4 specifies the research questions that this study aims to answer. Section 2.5 summarizes this chapter and reflects on its content.

[2] This chapter is partially based on: Gharesifard, M., Wehn, U., & van der Zaag, P. (2019b). What influences the establishment and functioning of community-based monitoring initiatives of water and environment? A conceptual framework. *Journal of Hydrology*, 579, 124033. doi:https://doi.org/10.1016/j.jhydrol.2019.124033

2.1 LITERATURE REVIEW

In order to develop the conceptual framework of this research, a large body of literature was critically reviewed. The focus of the review was on literature that has previously elicited relevant factors that affect the establishment and functioning of CBMs. This extensive literature review was conducted as an 'integrative review' (Bandara et al., 2015), which means it was informed by both theoretical and empirical evidence from the literature.

The reviewed literature was identified using three main methods; i) the author's knowledge about existing theoretical and empirical research in the field, ii) searching scientific databases (e.g. Web of Science, ScienceDirect, Scopus and Google Scholar), and iii) backward and forward snowballing (Van Wee & Banister, 2016). The literature review was started with a number of publications that were known to the researcher and which identified or mentioned influential factors on establishment, functioning or results of CBMs. Next, scientific databases were searched for additional relevant literature. The first step for defining the scope of the literature search was to identify different terminologies that are used to refer to CBM. As stated by Newman et al. (2011) the terminologies that refer to various citizen-based approaches in the field of Citizen Science still 'remains confusing' and there are a number of overlapping terms, which refer to the concept of CBM. Previous research (including Conrad & Hilchey, 2011; Kullenberg & Kasperowski, 2016; Newman et al., 2011; Whitelaw et al., 2003) has already identified and referred to these overlapping terminologies. For example, in a meta-analysis of Citizen Science literature Kullenberg and Kasperowski (2016) identified overlapping concepts such as 'community-based monitoring', 'volunteer monitoring' and 'participatory monitoring'. In another study, Newman et al. (2011) found an overlap between the terms 'community-based monitoring', 'citizen-based monitoring', 'collaborative monitoring' and 'volunteer monitoring'. In addition, as discussed in the Introduction Chapter, there is a close link between CBM and the concept of citizen observatories. This resulted in selecting the following set of terms for our literature search:

"community-based [environmental] monitoring", "participatory [environmental] monitoring", "collaborative [environmental] monitoring", "volunteer [environmental] monitoring", "citizen-based [environmental] monitoring", "Citizen observatory" and "Citizen observatories".

The idea of conducting a systematic search was tested. The aforementioned databases were searched for publications that referred to any of the identified terminologies in their title, abstract or keywords. The time span of the search was not limited and we included all documents that have ever been published in these repositories until 2018. This resulted in a very large number of publications (e.g. more than 8,000 records in Google Scholar). Reviewing this large number of documents was not manageable; therefore, the author considered filtering the records by adding keywords such as 'aspect', 'issue', 'dimension', 'factor', 'output', 'outcome' and 'impact'. This did not help reduce to the number of retrieved records to a great extent, mainly due to the fact that these terms are 'usual suspects' in a great majority of scientific publications. Therefore, backward and forward snowballing (Van Wee & Banister, 2016) was adopted as the main method for expanding the literature search results, i.e., where relevant, citations in or to the reviewed literature were identified and reviewed. In the case of backward snowballing, the process of finding additional literature included reviewing the reference list of already identified literature and scanning the abstract and conclusion sections of potentially relevant literature. Based on this initial assessment, newly identified literature was either excluded or marked for tentative inclusion. In some cases the researcher was interested to deepen his understanding about a specific concept that was identified in the reviewed references. Hence, forward snowballing was used, i.e. searching scientific databases to find the literature, which cited those references. Similarly, the abstract and conclusions of the identified publications were scanned, and if relevant, they were marked for potential inclusion. The final decision for inclusion or exclusion of the references, which were identified using backward and forward snowballing, was made after the full review of the references.

Besides their theoretical insights, the identified publications in the literature search contained empirical insights from a wide range of past and ongoing Citizen Science and CBMs. In addition, the researcher wanted to further complement the literature review with empirical evidence from a number of CBMs. There are an overwhelming number of CBM projects that could be included. For example, a recent inventory of Citizen Science and CBM projects by the European Commission includes 503 that have relevance for environmental policy (Bio Innovation Service, 2018). This list consists of a diverse set of projects with different thematic foci, including discontinued as well as ongoing and long-established projects. The researcher decided to examine the five citizen observatories of water and environment that had been funded under the 7th Framework Programme for Research and Technological Development in Europe (EU-FP7), namely WeSenseIt, CITI-SENSE, Citclops, COBWEB and OMNISCIENTIS. The main reason for this choice was the fact that these projects are considered 'pioneer' or 'legacy' CBM projects in Europe (EASME, 2016). This is due to the fact that these projects were the first attempt of the European Commission to 'demonstrate' the concept of 'Citizen Observatories' of the environment in Europe. These projects therefore produced insights about the setting up of several CBMs with diverse thematic foci in 16 countries in Europe and beyond (including the US and Israel). The experiences from these projects are now being applied in developing and scaling up several CBMs under the Horizon2020 funding program of the European Commission. It was also considered that these projects are relatively well-documented and a lot of information about these projects can be retrieved via publically accessible project reports. This yielded 67 project reports to the list of reviewed literature.

A thorough processing of these documents revealed that a great majority of the reports only focus on scientific or technical aspects of the projects and only 14 reports had references to aspects that influence the establishment and functioning of CBMs. This added a shortlist of 14 reports to the list of reviewed literature in this research.

The analytical method for reviewing the literature consisted of a mixed inductive and deductive approach. Starting inductively with no preconceived themes or categories, the researcher extracted different dimensions and aspects that had been identified in the reviewed literature and project reports as relevant factors for the establishment and functioning of Citizen Science or CBM initiatives. During the review process, the researcher found a close link between the reviewed literature and major theories on public participation in decision making as well as literature from the field of STS. Therefore, a number of relevant publications from those fields were deductively reviewed, which helped to further complement the researcher's understanding of the identified dimensions and aspects. Ultimately, the literature search and review resulted in identifying 70 publications, out of which 41 are peer-reviewed and 29 are non-peer reviewed documents including books, project reports and guidelines. It is important to mention that the literature search was conducted in English and therefore a language bias may have been introduced.

Next, the results of the literature review were clustered and synthesized. The process of clustering and synthesizing the findings included three overlapping and iterative steps. It started with coding the extracted aspects from the literature thematically. The emerging and recurrent themes were then (re)categorized and merged as the literature review progressed. This allowed the researcher to identify the prominent themes. These were then clustered and synthesized into the five dimensions that were generic across all CBMs, namely, 'goals and objectives', 'technology', 'participation', 'power dynamics' and 'results'. Each one of these dimensions consists of several aspects and raises a core question about the intrinsic nature of an initiative as well as the technological, institutional and political context in which it is embedded and with which it interacts. Table 2.1 summarizes the list of reviewed literature and indicates the relevance of the literature for the discussions in each dimension.

Table 2.1 Overview of the reviewed literature
Source:Gharesifard et al. (2019b)

Domain	Reference	Relevant dimension/aspect				
		Goals & Objectives	Technology	Participation	Power dynamics	Results
CBM, citizen observatories, broader field of citizen science and informal science education	(Bonney et al., 2009a)			✓		✓
	(Bonney et al., 2009b)					✓
	(Bonney et al., 2015)			✓		
	(Bonney et al., 2014)					✓
	(Brossard et al., 2005)					✓
	(Ciravegna et al., 2013)	✓	✓	✓	✓	
	(Conrad & Hilchey, 2011)				✓	✓
	(Conrad & Daoust, 2008)	✓		✓		
	(Cooper et al., 2007)	✓		✓		
	(Dickinson et al., 2012)			✓		
	(Dickinson et al., 2010)			✓		
	(Fernandez-Gimenez et al., 2008)					✓
	(Friedman, 2008)					✓
	(Gharesifard & Wehn, 2016a)	✓	✓	✓		
	(Gharesifard & Wehn, 2016b)	✓	✓			
	(Gharesifard et al., 2017)			✓	✓	
	(Gouveia & Fonseca, 2008)		✓		✓	
	(Haklay, 2015)			✓		
	(Irwin, 1995)			✓		
	(Irwin, 2001)				✓	
	(Irwin, 2015)	✓				
	(Jordan et al., 2012)					✓
	(Kieslinger et al., 2017)					✓
	(Kieslinger et al., 2018)					✓
	(Kimura & Kinchy, 2016)	✓				
	(Kullenberg & Kasperowski, 2016)			✓		
	(Liu et al., 2014)			✓		
	(National Research Council, 2009)					✓
	(Newman et al., 2012)		✓		✓	
	(Phillips et al., 2012)					✓
	(Phillips et al., 2014)					✓
	(Phillips et al., 2018)					✓
	(Pocock et al., 2014)			✓		
	(Roy et al., 2012)	✓		✓	✓	
	(Schäfer & Kieslinger, 2016)					✓
	(Shirk et al., 2012)			✓		✓
	(Silvertown, 2009)	✓				
	(Tredick et al., 2017)	✓				
	(Wehn et al., 2015a)			✓	✓	
	(Wehn & Almomani, 2019)	✓				
	(Wehn et al., 2017)				✓	✓
	(Wehn et al., 2015b)	✓	✓	✓	✓	
	(Wiggins et al., 2018)					✓
	(Wiggins & Crowston, 2011)				✓	
	(World Bank Group, 2016)	✓	✓		✓	✓

Domain	Reference	Relevant dimension/aspect				
		Goals & Objectives	Technology	Participation	Power dynamics	Results
Public participation in decision making	(Cleaver, 1999)	✓				
	(Fung, 2006)			✓	✓	
	(Kemerink et al., 2013)				✓	
	(Pahl-Wostl, 2009)				✓	
	(Macintosh, 2004)		✓	✓	✓	
	(Nelkin, 1975)				✓	
Science and Technology Studies (STS)	(Bijker et al., 2012)		✓			
	(MacKenzie & Wajcman, 1999)		✓			
	(Mansell & Wehn, 1998)		✓			
	(van Dijk, 2006)		✓			
	(Winner, 1986)		✓			
Project reports from EU-FP7 CBM projects	(Arpaci et al., 2016a)	✓				
	(Arpaci et al., 2016b)		✓			✓
	(Arpaci et al., 2015)	✓				
	(Bartonova et al., 2016)			✓		
	(COBWEB Consortium, 2015b)			✓	✓	
	(COBWEB Consortium, 2015a)	✓				
	(COBWEB Consortium, 2016)	✓				
	(COBWEB Consortium, 2017)	✓	✓		✓	✓
	(Gilardoni et al., 2013)	✓				
	(Novoa & Wernand, 2013)			✓		
	(OMNISCIENTIS Consortium, 2014)	✓	✓		✓	
	(Wehn de Montalvo et al., 2013)				✓	
	(Wehn et al., 2016)				✓	
	(WeSenseIt Consortium, 2016)		✓			

2.2 DIMENSIONS OF COMMUNITY-BASED MONITORING INITIATIVES OF WATER AND ENVIRONMENT

The following sections present the five generic dimensions of CBM that were identified in this research, namely, 'goals and objectives', 'technology', 'participation', 'power dynamics' and 'results'. The discussion in each section includes referencing the reviewed literature and clarifies how each dimension was informed by the literature review. Based on the discussions on each dimension, a core question is raised that aims to trigger critical thinking about that dimension.

2.2.1 Goals and objectives dimension of CBM initiatives

A large part of the reviewed literature either referred to examples of goals and objectives of CBM initiatives or highlighted the importance of studying these goals and objectives (e.g. Ciravegna et al., 2013; Conrad & Daoust, 2008; Gharesifard & Wehn, 2016a, 2016b;

Kimura & Kinchy, 2016; Roy et al., 2012; Tredick et al., 2017; Wehn et al., 2015b; World Bank Group, 2016). Kimura and Kinchy (2016) argue that without attention to the objectives of a Citizen Science initiative, it is not possible to study the quality and degree of participation in that initiative. The World Bank Group (2016) goes one step further and emphasizes that without understanding the 'goals and objectives' of digital citizen engagement (stated or otherwise), it is not possible to evaluate other dimensions of such initiatives.

CBM initiatives are typically formed around water or environment-related issues and their overarching objectives are often set by the project initiators and/or funders (Kimura & Kinchy, 2016). Much less frequently, objectives of CBM initiatives are co-defined in consultation with all, or a group of, concerned stakeholders. Collecting environmental data, raising environmental awareness, increasing public participation in monitoring and management of water or environment-related issues (e.g. flood risk management, water quality monitoring or environmental quality of public spaces), creating new forms of communication between citizens, scientists and decision makers, and developing enabling technologies for the aforementioned purposes are examples of the stated objectives of the EU-FP7 CBM projects that we reviewed in this study (Arpaci et al., 2016a; COBWEB Consortium, 2015a, 2016; Gilardoni et al., 2013; OMNISCIENTIS Consortium, 2014).

No matter what the objectives of an initiative are, or how they are defined, it is highly important to realize the different stakeholders may have divergent interests or values (Silvertown, 2009; Wehn & Almomani, 2019) and the objectives of a CBM initiative may incline more towards preferences and wishes of some stakeholders than those of others. Moreover, the overarching goals of a CBM initiative may not be bias-free and might have been influenced by vested interests of funders, researchers and technology providers. This may result in some stakeholders benefiting more from the initiative than others; an issue that is among the lessons learned from the CITI-SENSE projects (Arpaci et al., 2016a). Previous research efforts have mainly focused on overarching objectives of the CBM initiatives and do not investigate how actor-specific goals might differ from one actor (group) to another (Gharesifard & Wehn, 2016b). There is also little attention to the synergies and contradictions between these goals and the overarching objectives of a CBM initiative. A thorough study of the CBM objectives and goals of the actors involved may help answer a number of questions that are often mentioned with regards to paradoxes of participatory processes (Cleaver, 1999); for example, who gains most from the CBM activities and whose interest is least reflected.

The initial objectives of a CBM initiative may be modified or evolve over time (Cooper et al., 2007; Irwin, 2015). This can happen because of different reasons, for example, financial or technological constraints, power dynamics between involved stakeholders, or adjustment of project ambitions. The latter happened to both the CITI-SENSE and the

COBWEB projects which reportedly changed their objectives due to their realization of challenges with achieving their initial objectives (Arpaci et al., 2016a; COBWEB Consortium, 2017). It is therefore also important to monitor any changes in the objectives and to understand why such changes have happened. Monitoring the objectives and the extent of their achievement is a benchmark for assessing intended outcomes and impacts of a CBM.

Thus, the first core question is *'What are the overarching objectives and actor-specific goals of the CBM initiative and to what extent does the design of the initiative help achieve those goals/ objectives?'.*

2.2.2 Technology dimension of CBM initiatives

Recent technological developments and advancements in ICTs have transformed and accelerated CBM initiatives to a great extent and have created new possibilities for stakeholder participation in science and policy. It is therefore important to study how to engage different stakeholders in a CBM initiative (Ciravegna et al., 2013; Gharesifard & Wehn, 2016a, 2016b; Macintosh, 2004; Wehn et al., 2015b; World Bank Group, 2016).

The technological choices for a CBM initiative can be broadly divided into two main categories; those technologies that existed and were being used before the establishment of the initiative and those that are newly developed or introduced to create the desired functionalities for a CBM initiative. Each of these categories can be further broken down based on the functionality that is envisioned for them in a CBM initiative (e.g. data collection, visualization, or communication). Adopting both existing and newly introduced technologies comes with a number of social, political, economic and even cultural changes that needs to be carefully considered before selecting different technological options. This can be done by asking who is being included or excluded (intentionally or unintentionally) as a result of specific technological choices; the extent to which these choices create and/or maintain specific social conditions that favor some and marginalize others; and the degree to which they are compatible with (internal or external) social and political structures and relationships. Therefore, while establishing a CBM initiative, there is a need for gaining an understanding of existing infrastructure and availability of different forms of access to a wide range of possible technologies by different stakeholder groups (Gouveia & Fonseca, 2008; Newman et al., 2012). This closely relates to discussions on the digital divide and forms of access to technology (material, motivational, usage, and skills) (e.g. van Dijk, 2006) and helps identify included/excluded groups resulting from choices of technology.

Moreover, as many STS scholars have emphasized, the perception that technology changes solely as a result of scientific advancements or on its own accord is a passive way of conceiving technology that focuses on how to adapt to technological changes rather than how they shape, or are shaped by society, the economy or politics (Bijker et al., 2012; MacKenzie & Wajcman, 1999; Mansell & Wehn, 1998; Winner, 1986). For example, a web-platform of a CBM initiative that focuses on the issue of water or air quality is not only shaped by technological components that are a prerequisite for its creation and functioning, but also, for example, by vested interests or economic constraints of its developers or the end users.

In the case of the five EU-FP7 projects that were reviewed for this study, most of the projects started their CBM activities with a technology-driven model; meaning they started developing tools and technologies that seemed suitable for achieving their objectives without gaining a thorough understanding of the social, economic and political system that they were operating in. For example COBWEB mentions in its final report that they entered 'a process of rapid prototyping software development' based on their 'identified requirements' and only when they reached a certain level of technology readiness, started to engage citizens (COBWEB Consortium, 2017). The interpretation of the results of this approach is different among different projects; some call it a 'great success' (e.g. COBWEB Consortium, 2017; OMNISCIENTIS Consortium, 2014) and some identify it as a source of 'major difficulties' in developing the CBM initiatives (e.g. Arpaci et al., 2016b; Arpaci et al., 2015; WeSenseIt Consortium, 2016). For example, Arpaci et al. (2016b) mentioned that technological developments became one of the 'key motives' of the CITI-SENSE project and this caused major difficulties with engaging and empowering citizens because it shifted the project's attention and resources away from engagement and empowerment activities.

This dimension therefore considers how enabling technologies of a CBM initiative have been shaped and how these relate to existing infrastructure as well as social and technological capabilities, by asking *'How effective and appropriate are the choices and delivery of the selected technologies?'*

2.2.3 Participation dimension of CBM initiatives

Enhancing public participation in environmental monitoring, planning or management is a core concept of CBM initiatives and the first step to understanding the state of change in such participation processes is to deepen our understanding of the existing participation dynamics related to the water or environmental issue in focus.

Since the level of engagement and commitment of participants differs across different CBM initiatives, at different stages of its development and across different actors, it is important to first clarify what 'participation' in a CBM initiative implies. Participation in a CBM initiative is closely related to the higher purpose that the initiative serves and the type of activities that are being conducted as part of it. There have been a number of attempts to classify Citizen Science initiatives and the extent of participation in these initiatives. For example Bonney et al. (2009a) classified Citizen Science projects based on the degree of participants' involvement in scientific investigations steps into 'contributory', 'collaborative' and 'co-created projects'. Shirk et al. (2012) expanded these categories to five models of public participation in scientific research that included 'contractual', 'contributory', 'collaborative', 'co-created' and 'collegial'. Haklay (2015) distinguished between six levels of participation based on the extent of engagement and commitment of participants, namely; 'passive sensing', 'volunteer computing', 'volunteer thinking', 'environmental and ecological observations', 'participatory sensing' and 'civic/community science'. Based on the roles that citizens can take on, and the higher aim of CBM, Wehn et al. (2015a) distinguished between initiatives that aim for environmental monitoring, co-operative planning, and environmental stewardship. Another example is a meta-analysis of Citizen Science literature in which Kullenberg and Kasperowski (2016) identified three types of initiatives according to their higher purpose, namely Citizen Science as 'a method', 'public engagement with science and policy', or 'civic mobilization'. The purpose of highlighting these typologies is not to prescribe or recommend an existing typology; rather, it is to emphasis the importance of considering different typologies of participation during the evaluation of CBM initiatives and clarifying what participation in an initiative actually entails.

Previous studies have identified 'geographic scope' as an aspect that is often linked to the issue in focus of the CBM initiative and shows its breadth of focus (Cooper et al., 2007; Haklay, 2015; Macintosh, 2004; Roy et al., 2012; Wehn et al., 2015b). Moreover, the geographic scope of a CBM initiative determines the spectrum of currently involved and affected stakeholders and helps identify the potential pool of participants in the initiative. Moreover, the geographic scope of a CBM may change over time, for example, as a result of its growth or its change of focus.

'Participant groups' are the actors or stakeholders who are involved in a CBM initiative (Ciravegna et al., 2013; Conrad & Daoust, 2008; Macintosh, 2004; Wehn et al., 2015b). Depending on the type and objectives of an initiative, these are normally individuals, groups or organizations who had a role in the design and setting up of the initiative, are actively involved in the initiative via data collection/sharing, analysis, aggregation and visualization, and/or use its outputs for improving policy or decision making processes. It is also equally important to understand which groups of stakeholders are not represented in a CBM initiative and to critically reflect on consequences of their absence. Although contributing to public policy and decision making processes is far from easy (Irwin, 1995), and may not be among the objectives of a CBM initiative, studying the stakeholders already involved in decision making processes offers the possibility to know who might be interested in the data and knowledge generated via CBM, and also helps deepen our understanding of included and excluded groups (Wehn et al., 2015b). This is especially necessary for CBMs that (have the ambition to) move beyond mere data collection and perceive a more active role for citizens in influencing decision making processes (Bonney et al., 2015; Dickinson et al., 2012; Kullenberg & Kasperowski, 2016).

'Effort required to participate' and 'support offered for participation' are two other aspects of the participation dimension that have been identified by previous research (Ciravegna et al., 2013; Conrad & Daoust, 2008; Dickinson et al., 2010; Gharesifard & Wehn, 2016a; Gharesifard et al., 2017; Liu et al., 2014; Pocock et al., 2014; Roy et al., 2012; Rutten et al., 2017). 'Effort required to participate' refers to different types of requirements and investments that are needed from participants such as time or monetary investments, or expertise. CITISENSE, COBWEB and Citclops identified examples of knowledge requirements such as participants' understanding of complex environmental issues, e.g. air pollution or flooding, and their experience with data collection processes as factors that influenced their project engagement efforts (Bartonova et al., 2016; COBWEB Consortium, 2015b; Novoa & Wernand, 2013). 'Support offered for participation' considers the investments made by the initiators to communicate about the CBM initiative and to facilitate public participation via, for example, flexible participation methods, easy to use web-platforms and mobile applications, incentives provided for participation, and the availability of supporting materials, guidelines and trainings.

Communication in the context of a CBM initiative can go beyond just 'data push' and in many cases CBMs act as a medium for facilitating communication between different stakeholders (Liu et al., 2014; Wehn et al., 2015b). Identifying the existing communication channels and current patterns of information flow between different stakeholders before the establishment of such an initiative is essential for understanding existing norms and mental frameworks for communication, and helps explain how an initiative has affected these interaction patterns. Ciravegna et al. (2013) and Wehn et al. (2015b) distinguished between three different patterns of information flow, namely 'unidirectional', 'bi-directional' and 'interactive'. 'Pattern of communication' is considered to distinguish between CBMs that only act as recipient of the data and those initiatives that either provide feedback through different communication channels or form an interactive exchange of information among the triangle of citizens, data aggregators and policy makers (Wehn et al., 2015a) that may alter the existing pattern of information flow between these stakeholders.

Finally, it is important to understand how different stakeholders participate in a CBM initiative. Studies in the field of public participation in decision making provide insights about methods of participation in public settings. For example, Fung (2006) identified six modes of communication (i.e. 'listen as spectator', 'express preferences', and 'develop preferences') and decision making (i.e. 'aggregate and bargain', 'deliberate and negotiate', and 'technical expertise') and defined it as the way by which "participants interact within a venue of public discussion or decision" (Fung, 2006, p. 68). Wehn et al. (2015b) adjusted these modes for CBMs by adding implicit and explicit data collection to this spectrum. Analyzing these methods of communication and participation in decision making before and after the initiation of a CBM helps to depict how participants used to interact in public discussions or decisions on the water or environment-related issue in focus of the CBM initiative and how the initiative may have mediated or altered these interactions (Wehn et al., 2015b).

Studying the identified aspects in this dimensions helps to understand the participation dynamics in a CBM initiative and answer the key question *'Who participates in the CBM initiative and how, and who does not?'*

2.2.4 Power dynamics dimension of CBM initiatives

Environmental governance is inherently a political process that involves competing interests and conflicting norms and values for different actors (Cleaver, 1999). Since a CBM is established to help better understand or address a specific water or environment-related issue, existing power dynamics related to the governance of those issues are inevitably involved (Newman et al., 2012; Wehn et al., 2015b). Furthermore, the power (im)balance among different actors in a CBM creates internal power dynamics that shape the objectives, functioning and outcomes of the initiatives. This section summarizes the results of our review regarding internal and external power dynamics of a CBM initiative.

The importance of understanding the social, institutional and political context in which a CBM operates has been highlighted in previous research (e.g. Emmett Environmental Law and Policy Clinic, 2019; Irwin, 2001; Wehn et al., 2017). Institutions here refer to the multi-level social and legal arrangements that regulate actors' behavior (Pahl-Wostl, 2009); policies are understood as implicit and explicit procedures in managing 'natural resources and infrastructure' (Kemerink et al., 2013). Understanding these contextual realities helps depict how the current system of decision making works by providing insights on the formal and informal rules and regulations related to the issue in focus of the CBM initiative; the extent to which these rules and regulations are being implemented and enforced; the roles and responsibilities of different actors; and the role of public participation in these processes. Both the WeSenseIt and COBWEB projects highlighted the importance of understanding the institutional and political context. The water governance context of the case studies in WeSenseIt were systematically analyzed and reported in a number of project reports and publications (e.g. Wehn de Montalvo et al., 2013; Wehn et al., 2016; Wehn et al., 2015b). In addition, one of the lessons learned from COBWEB is that without understanding the concepts and processes that underpin decision making processes, it is not possible to understand the value added of the data produced by a CBM (COBWEB Consortium, 2015b).

'Authority and power' or the actual level of impact of different stakeholders on decision making processes related to the environmental issue in focus of the CBM is the next aspect of power dynamics. This aspect focuses on a dilemma that Nelkin described as follows: "the complexity of public decisions seems to require highly specialized and esoteric knowledge, and those who control this knowledge have considerable power. Yet democratic ideology suggests that people must be able to influence policy decisions that affect their lives" (Nelkin, 1975, p. 37). CBM initiatives have the potential to close, or at least narrow, this knowledge gap and reduce a power imbalance in decision making processes, but this is not definitive. A lot of CBM projects claim that they have increased citizens' influence on decision making processes, but this claim is often hypothetical, not evidence-based, or at least not well documented. For example, without providing details, the OMNISCIENTIS final report claims that 'local environmental governance' was enhanced through citizen participation in project activities and monthly meetings (OMNISCIENTIS Consortium, 2014). In another example, the final report of the COBWEB identifies increased citizens' influence on environmental governance as its 'chief expected impact', but there is no reported evidence on how this impact actually happened (COBWEB Consortium, 2017). However, a thorough study of the water governance aspects in the WeSenseIt project revealed that it is very difficult, if not impossible, to capture changes in citizens' authority and power during the lifetime of a project (Wehn et al., 2016). Thus it is important to investigate the levels of authority and power of different stakeholders before and long after establishment of a CBM. The five different levels of authority and power suggested by Fung (2006) can serve to assess this aspect (Wehn et al., 2015b). These levels start from individual education and increase to communicative influence, advise/consult, co-govern, and finally direct authority.

Another aspect to be considered while examining power dynamics is 'access to and control over data' (Emmett Environmental Law and Policy Clinic, 2019; Gouveia & Fonseca, 2008; Irwin, 2001; Roy et al., 2012; Wehn et al., 2016; World Bank Group, 2016). Ownership of the data and the ability to analyze the data are directly linked to its actual use. Who defines the level of access to water or environment-related data for different participants? Who decides on the quality control procedure? Who has the required skills to analyze the data? And who can veto the data collection and aggregation procedures and the publication of 'harmful' data? These are the type of questions that may be asked to determine the access to and control over data before and after the establishment of a CBM.

The stakeholders who establish a CBM usually have a strong say in defining its overarching objectives, governance structure, participation mechanisms and the chosen technologies. The 'establishment mechanism' is described as the way in which the CBM initiative is founded and has four distinct types; 'top-down', 'bottom-up', 'commerce driven', and 'co-created'. The first two types were previously identified by Ciravegna et al. (2013) and (using different titles, i.e. 'consultative' and 'transformative') by Conrad and Hilchey (2011); for top-down systems, authorities and stakeholders at higher levels of policy or decision making initiate the CBM while for bottom-up systems, stakeholders such as citizens or volunteers are the initiators. The commerce-driven model is added to capture the establishment mechanism of those observatories that have been set up neither by official administrative bodies nor by lower levels of decision making hierarchy but are market-based and for-profit (Gharesifard et al., 2017). Finally, the co-created, collaborative or co-designed CBM is a novel approach that aims to provide to as many interested stakeholders as possible a chance to influence the design and functioning of CBMs by involving them in different steps of its establishment process (Conrad & Hilchey, 2011; Wehn et al., 2015a).

'Revenue stream to sustain the initiative' is the aspect that depicts how a CBM generates its revenue or receives its required funding. This helps to explain critical issues such as financial motivations behind running the CBM, its sustainability, data ownership and the level of access to the generated information for the general public. Despite its importance, this aspect has not received much attention in previous research. For example, the EU-FP7 CBM projects that we reviewed in this study were pilot CBM projects and therefore did not consider revenue streams for sustaining the initiatives that they established. Macintosh (2004) touches upon funding issues when she discusses 'resources and promotion' and briefly mentions that due to the novelty of e-participation initiatives, they are mostly funded by national governments through their R&D budget. Wiggins and Crowston (2011) mentioned that the largest Citizen Science projects of the US National Science Foundation (NSF) have received their funding in the form of sponsorships, sales referrals, or licensing. Perhaps the most comprehensive categorization of potential revenue streams of CBMs is the seven categories identified by Gharesifard et al. (2017). This classification is adopted from Osterwalder and Pigneur (2010) and further adjusted to best capture the revenue streams for a CBM. This classification includes 'government sponsorship', 'data/information usage fee', 'subscription fee', 'asset sale', 'advertising', 'licensing' and 'donation'.

A critical analysis of the aspects that were introduced in this dimension will increase our understanding of *'who controls and influences the CBM initiative and how?'* This question helps depict both internal and external power dynamics of a CBM.

2.2.5 Results dimension

During recent years, the expertise to set up and run Citizen Science projects and CBM initiatives has grown at a very fast pace, but as Phillips et al. (2014) stated, there is still a 'critical gap' between these competencies and those required to evaluate the value and impact of these initiatives. A considerable part of the reviewed literature in this research focuses on evaluating Citizen Science and CBM projects, public participation in scientific research, science learning in informal environments and e-participation (Bonney et al., 2009a; Bonney et al., 2009b; Bonney et al., 2014; Brossard et al., 2005; Fernandez-Gimenez et al., 2008; Friedman, 2008; Jordan et al., 2012; Kieslinger et al., 2017, 2018; National Research Council, 2009; Phillips et al., 2012; Phillips et al., 2014; Phillips et al.,

2018; Schäfer & Kieslinger, 2016; Shirk et al., 2012; Wehn et al., 2017; Wiggins et al., 2018; World Bank Group, 2016). Although providing detailed instructions on how to evaluate CBM projects is beyond the scope of this doctoral research, the review of this literature resulted in a number of critical points that need to be considered while evaluating CBM initiatives.

Most of the reviewed literature, including Brossard et al. (2005), National Research Council (2009), Bonney et al. (2009a), Bonney et al. (2009b), Jordan et al. (2012), Bonney et al. (2014), Schäfer and Kieslinger (2016), Kieslinger et al. (2017), Wiggins et al. (2018) and Kieslinger et al. (2018), does not distinguish between outputs (i.e. direct products) of an initiative, and its outcomes and impacts (i.e. the short term, mid-term and long-term changes) that can be attributed to the initiative. Friedman (2008) proposed the use of the Logic Model for evaluating Informal Science Education programs and distinguishing between outputs (e.g. the number of participants in these programs), outcomes (e.g. improved understanding about a certain topic among the participants) and impacts (e.g. lasting changes in the behavior of participants). This idea was later adopted by a number of researchers for evaluating the results of Citizen Science projects (e.g. Phillips et al., 2012; Phillips et al., 2014; Phillips et al., 2018; Shirk et al., 2012; Wehn et al., 2017). It is important to distinguish between the direct products of a CBM and the changes (e.g. in the status quo of water resources) that can be attributed to the existence and functioning of the initiative. Failing to recognize or acknowledge this difference or the fact that the results of CBMs are likely to evolve and change over time may have practical implications for evaluation processes. Evaluating the direct products of an initiative may require fundamentally different methods than those required to assess the changes that result from that initiative. For the purpose of this research, we define 'outputs' as the direct products of a CBM initiative; 'outcomes' as incidental and short-term changes that can be attributed to the existence of the CBM initiative; and, 'impacts' as long-term outcomes that are broad in scope and are associated with structural changes.

The review also revealed that most of the existing literature predominantly focuses on scientific outputs, individual learning outcomes and to some extent on societal outcomes of an initiative (e.g. Bonney et al., 2009a; Bonney et al., 2009b; Friedman, 2008; National Research Council, 2009; Phillips et al., 2012; Phillips et al., 2014; Phillips et al., 2018). Broader outcome categories, such as environmental, economic and governance-related outcomes of the initiatives (e.g. change in policies, legislations or actors' authority), are often ignored, assumed or speculated. In a review of 10 years of relevant Citizen Science literature, Conrad and Hilchey (2011) also concluded that the environment and governance-related success stories of CBMs are largely undocumented. This is mainly because environmental, economic and governance-related changes are complex in nature, interrelated, difficult to study and unfold over a long period of time. In addition, at least for funded CBM projects, evaluation often happens partially, superficially, towards the end of the project and primarily for reporting purposes. As soon as the project funding ends, hardly any effort is made to evaluate its mid-term and long-term impacts. For example, the final report of the CITI-SENSE project indicates that it was not possible to access the wider societal impacts of the project (Arpaci et al., 2016b). In another example, the chief expected impact from COBWEB was "enabling greater citizen influence in environmental governance" (COBWEB Consortium, 2017, p. 4), however, this hypothesized output is not supported by evidence. Studies such as Fernandez-Gimenez et al. (2008), Shirk et al. (2012), Bonney et al. (2014), World Bank Group (2016), Kieslinger et al. (2017), Wehn et al. (2017) and Wiggins et al. (2018) consider broader social, environmental, economic and governance-related outputs and outcomes of CBMs. For the purpose of this research, we combined the typologies of outputs, outcomes and impacts from the reviewed literature and classified those into six meta-categories. Table 2.2 provides an overview of these categories and sample outputs, outcomes and impacts of a CBM.

Table 2.2 Meta-categories of results and examples of outputs, outcomes and impacts of
CBM initiatives
Source:Gharesifard et al. (2019b)

Meta-categories of CBM results	Examples of outputs	Examples of outcomes and impacts
Individual	• Publicly accessible water or environment-related databases/ datasets • Improved individual knowledge or understanding of an environmental issue • Networks of like-minded people • Development of new skill	• Improved sense of place and/or stewardship • Improved relationship between science, society and authorities
Scientific	• Datasets and information about a topic of interest • Scientific publications	• Advancement in scientific understandings about a topic • Improved relationship between science, society and authorities
Societal	• Publicly accessible water or environment-related databases/datasets • Improved average level of knowledge or understanding of an environmental issue	• Increased social capital • Improved average level of health within society • Improved livelihoods • Improved relationship between science, society and authorities
Economic	• Immediate value of newly produced datasets (e.g. for scientists or authorities) • Creation of new jobs	• Reduced costs of resource monitoring and management • Long-term economic return of the initiative's products and services
Environmental	• Improved knowledge or understanding of an environment issue • Higher level of awareness about, and responsibility for protection of natural resources	• Improved protection of natural resources • Improved status quo of water resources or the environment
Governance	• New channels of communication between decision makers, scientists and citizens • Additional information about the environment and natural resources	• Better-informed water or environment-related decisions & policy • Change in legislations or processes of decision making about natural resources management • Change in balance of power in decision making processes

The researcher acknowledges the interdependencies and overlaps between these meta-categories of results, and believes that they cannot be studied independent from one another. This is demonstrated by providing examples of outputs and outcomes in Table 2.2 that may belong to more than one category of results. However, recognizing these meta-categories of results will help guide our conceptual thinking on the design and implementation of evaluation processes and it will help communicating the outputs, outcomes and impacts with those who may be interested in a specific domain of results (e.g. scientists or water managers). The researcher also believes that not all six meta-categories of results may be relevant for each CBM since the results of an initiative, among other things, are related to its objectives.

The last question is thus *'what are the expected and realized outputs, outcomes & impacts of the CBM initiative?'* It is important to note that outputs, outcomes and impacts can be positive or negative. Furthermore, comparing outputs, outcomes and impacts with the objectives of a CBM will help identify intended and unintended results.

2.3 SYNTHESIS OF THE CONCEPTUAL FRAMEWORK

The proposed conceptual framework for examining the Contextual setting, Process evaluation and Impact assessment of a CBM initiative (in short the CPI Framework) is illustrated in Figure 2.1. Different aspects of each dimension are categorized and marked based on their relevance for the issue/context in focus of a CBM initiative, and whether they are internal to the initiative (see the legend on the lower-left of the framework). The aspects discussed in the two dimensions 'goals & objectives' and 'results' are directly related to the CBM, while the 'technology', 'participation', and 'power dynamics' dimensions cover a wide range of contextual factors related to the issue in focus of the initiative (e.g. the institutional and political context), its internal dynamics (e.g. its establishment mechanism), or both (e.g. communication paradigm about the issue, and within the CBM initiative). The circular shape of the CPI Framework is meant to acknowledge and emphasize the interdependencies between the five dimensions.

The distinction made between the context-related and initiative-related aspects broadens the applicability of the CPI Framework for different purposes and makes it suitable for context analysis, process evaluation and impact assessment of a CBM. For the purpose of context analysis, only the aspects that focus on the issue (marked with a circle or circle & square sign in the framework) need to be examined. For impact assessment, the classifications provided for outputs, outcomes and impacts can guide the evaluation of results generated by a CBM initiative. For process evaluation purposes, the core questions raised by the CPI Framework (e.g. who participates in the CBM initiative and how, and who does not? or who controls and influences the CBM initiative and how?) serve to analyze the processes that led to the outputs, outcomes and impacts of the CBMs and generate insights into why and how positive, negative, intended and unintended results are (not) being achieved.

In order to be able to capture changes in the contextual setting and to describe the establishment, functioning and results of a CBM, it is necessary to study these dimensions and aspects at different points in time; at least once at the beginning of the establishment process of an initiative and once at the time of evaluating its outputs, outcomes and impacts.

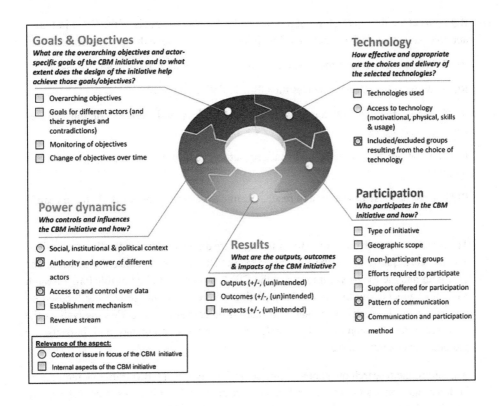

Figure 2.1 The CPI Framework
Source:Gharesifard et al. (2019b)

2.4 RESEARCH QUESTIONS

In line with the overarching and specific objectives of this research that were introduced in Chapter 1, and the introduced conceptual framework in the current chapter, this doctoral research focused on answering the following six research questions (RQ.1 to R.Q.6) for two CBM initiatives of the Ground Truth 2.0 project. These CBM initiatives and the logic for their selection are introduced in the next chapter.

RQ.1: What are the overarching objectives and actor-specific goals of the CBM initiative and to what extent does the design of the initiative help achieve those goals/objectives?

RQ.2: Who participates in the CBM initiative and how, and who does not?

RQ.3: Who controls and influences the CBM initiative and how?

RQ.4: How effective and appropriate are the choices and delivery of the selected technologies?

RQ.5: What are the expected and realized outputs and interim outcomes of the CBM initiative?

RQ.6: How do these processes and results change over time (approximately three years)?

The first five research questions (RQ.1 to RQ.5) are directly related to the five dimensions and the core questions of the CPI Framework. Due to the fact that the some of the outcomes and the long-term impacts of the CBM initiatives studied of this PhD research have not happened by the time of completing this research, the researcher focused on understanding the realized outputs and interim outcomes of the CBMs, as well as the expected future outcomes and impacts. These outputs, outcomes and impacts are presented in chapter 4 and 5.

Answering the research questions of this study for any given CBM initiative requires an in-depth understanding of several context-related and initiative-related aspects. In addition, due to the dynamic and evolving nature of the CBM initiatives and the context in which they are operating, there is a need for a research question that helps capturing the changes in the status of the five aforementioned dimensions over time. Therefore, a sixth research question (RQ.6) is introduced to help capture these changes over time.

The research matrix provided in Figure 2.2 illustrates the link between the six research questions and the specific research objectives that were introduced in Chapter 1 of this dissertation.

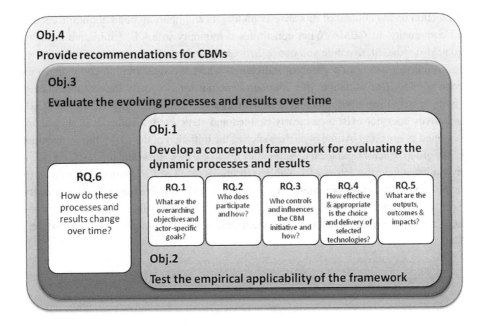

Figure 2.2 Research matrix

It is important to clarify that this doctoral research has been conducted in the context the Ground Truth 2.0 project and uses a case study approach. Therefore, the above-mentioned research questions have been analyzed in the context of two case studies within this project. The case studies and the rationale for their selection are explained in the next chapter.

2.5 SUMMARY AND REFLECTIONS

The development of the CPI Framework was in line with the first objective of this doctoral research, i.e. to develop a conceptual framework for evaluating the factors that influence the establishment, functioning and outcomes of community-based monitoring initiatives of water and environment. The current chapter introduced the CPI Framework and the steps, which were taken for its development. This framework consists of five dimensions that can help unpack what influences the establishment and functioning of a CBM initiative. It was demonstrated that the CPI Framework builds on an extensive review of the literature and theories in the field of citizen science and other affiliated fields of research. Finally, this chapter introduced the research questions of this study and showed their link to the core questions raised by the CPI Framework, and the objectives of the research.

Reflection on the content of this chapter highlights an important point about the concept of 'community' in CBMs. What constitutes community in a CBM initiative is highly context-dependent, dynamic and hence difficult to analyze. In general, a community is a social unit consisting of a group of individuals who have something in common. In the context of a CBM initiative, this common denominator is usually interest in, concerns about, or stake in an environmental issue. However, the community in the context of CBM initiatives does not exist as a clearly defined and static entity. Rather it is shaped and reshaped at any point in time by factors such as the initiation of the CBM, the composition of the group of actors and local communities involved, their goals and interests, the power dynamics among the actors and enabling technologies. The CPI Framework allows for understanding how the community in a particular CBM initiative has evolved in practice. Such a framework provides an interpretation of the community that it is used to evaluate, and therefore helps construct or define the meaning of community in a CBM initiative.

3

METHODOLOGY[3]

Having introduced the conceptual framework and the research questions in Chapter 2, the present chapter explains the methodological choices and steps that were taken towards applying the conceptual framework of the research for answering the research questions. Section 3.1 describes the research strategy for conducting this study, notably the choice of a case study approach, section of the case studies and the rationale for a two phase empirical research. Section 3.2 is dedicated to introducing the sources of data that informed the empirical research in this study. The methodologies used for data collection during interviews in the two phases of empirical research are explained in sections 3.3 and 3.4. Data analysis methods are presented and discussed in Section 3.5. Section 3.6 summarizes the research design by clarifying the link between the steps taken for conducting this research. This chapter is concluded in section 3.7 by outlining the main methodological steps and choices.

---

[3] This chapter is partially based on the following publications:

Gharesifard, M., Wehn, U., & van der Zaag, P. (2019a). Context matters: a baseline analysis of contextual realities for two community-based monitoring initiatives of water and environment in Europe and Africa. *Journal of Hydrology*, 124144. doi:https://doi.org/10.1016/j.jhydrol.2019.124144

Gharesifard, M., Wehn, U., & van der Zaag, P. (2019b). What influences the establishment and functioning of community-based monitoring initiatives of water and environment? A conceptual framework. *Journal of Hydrology*, 579, 124033. doi:https://doi.org/10.1016/j.jhydrol.2019.124033

3.1 Research strategy

3.1.1 Case study approach

This research employed a case study approach, which is a commonly applied method in social science studies. One of the most widely referred to definitions of the case study approach describes it as "an empirical inquiry that investigates a contemporary phenomenon within its real-life context when the boundaries between phenomenon and context are not clearly evident and in which multiple sources of evidence are used" (Yin, 1984, p. 23). As this definition suggests, the case study approach is normally used when a researcher aims to gain in-depth understanding of a complex reality. This doctoral study aimed to understand the factors that affect the establishment and outcomes of a CBM initiative, which is an example of both a contemporary phenomenon and a complex reality. CBM initiatives do not operate in a void; rather they are embedded in the social, environmental and economic context in which they are being established. In order to be able to study such a complex system, an enabling strategy was required to allow the researcher to study this phenomenon within its real life contexts (i.e. the social, environmental, economic, and political settings and realities that exist in a case study) and therefore the case study approach is chosen for this research.

3.1.2 Selection of case studies

This doctoral research was carried out between May 2015 and November 2019 and was funded by both the WeSenseIt[4] and Ground Truth 2.0[5] projects. The proposal of this research was written during the lifetime of WeSenseIt, while the empirical research was carried out during the funding period of Ground Truth 2.0. Hence, the case studies of this doctoral research were selected from the six case studies of the Ground Truth 2.0 project, which included four European (the Netherlands, Spain, Belgium and Sweden) and two African (Kenya and Zambia) cases.

Due to the fact that conducting an in-depth empirical research within all six case studies of the Ground Truth 2.0 project was not feasible, the researcher selected two case studies from this list, namely the Dutch and the Kenyan cases. A number of criteria were considered for making this selection. These considerations are explained hereafter.

For the purpose of empirical data collection, factors such as accessibility of the cases (in terms of geographic location), language of the cases and the operational condition of the

---

[4] WeSenseIt received funding from the EU's Seventh Framework Programme for Research (FP7), started at October 2012 and ended at September 2016. For more information see:
https://cordis.europa.eu/project/rcn/106532/factsheet/en
[5] Ground Truth 2.0 received funding from the EU's Horizon 2020 program, started at September 2016 and ended at December 2019. For more information see: https://cordis.europa.eu/project/rcn/203387/factsheet/en

CBM initiatives were considered. European countries with a higher English Proficiency Index (EF Education First, 2015) were given preference because the researcher was more familiar with this language than those in the other case study locations. The Netherlands received the highest overall score among the European cases, because the researcher was living there and the country had the second highest English proficiency rank in Europe (EF Education First, 2015). Moreover, the CBM initiative in the Netherlands case focused on the issue of pluvial flooding, which is aligned with the researcher's area of expertise i.e. the field of water management.

For cross-case comparison purposes, the researcher wanted to include an African case study in this research. The idea was to allow a cross-case compression of the process of establishment and functioning of a CBM initiative in the context of a 'developed' versus a 'developing' country. Compared to the European cases, both Kenya and Zambia were much less accessible because of their distance from the Netherlands and they both had the advantage of having English as their official language. However, at the time of making this selection, the CBM initiative in the Zambia case was still at a very preliminary stage of its development and therefore the Kenyan case was selected as the second case study for this research.

3.1.3 The rationale for a two phase research

The main objective of this research was to conduct a systematic evaluation of the factors that influence the establishment, functioning and outcomes of community-based monitoring initiatives. These factors could be intrinsic to a CBM initiative or related to the contextual realities in which the CBM is being embedded, and with which it interacts. Therefore, the researcher designed a two-phase research, in which phase 1 was dedicated to understanding the contextual settings in which the two CBM initiatives were being established, and phase 2 focused on understanding the intrinsic factors, as well as the outputs, outcomes and impacts of these two CBMs. In terms of timing, the empirical research in phase 1 was designed and conducted at the early stage of the development of the two CBMs in 2016 and 2017, while phase 2 of the research was designed and conducted towards the end of the Ground Truth 2.0 project in 2019.

3.2 SOURCES OF DATA

The research reported in this doctoral dissertation is a qualitative empirical research that is informed by both primary and secondary data.

The primary sources of data consist of the data collected by the researcher during the interviews in the two aforementioned phases of research, as well as the observations made during stakeholder meetings (i.e. co-design and planning meetings) and field visits in the two case studies by the researcher or other members of the GT2.0 team. In total 92 in-depth interviews were conducted in the two phases of this research. Table 3.1 summarizes the number of interviews in the two phases and across the case studies.

Table 3.1 Number of interviews in the two phases and across the case studies

	The Netherlands case	Kenya case
Number of interviews in Phase 1	26	34
Number of interviews in Phase 2	12	20
Total number of interviews in Phase 1 & 2	38	54

Observations were made during eleven stakeholder meetings and field visits in the two cases; eight in the Dutch case and three in the Kenyan case. In order to document the observations made during these events, an observation protocol was developed and used that allowed for documenting factors such as the goals of the meeting, physical surroundings, facilitation of the session, interactions between participants and direct quotes (For details see Annex 1). Moreover, for project purposes, a logbook of each event was filled by project team members and facilitators that included overlapping and additional observations. For example, these logbooks recorded information such as the purpose of the meeting, participants in the meeting, interactions between participants and lessons learned from the meeting. The observation protocols and logbooks were filled in during or shortly after the end of each meeting. In addition, the researcher was a member of the Kenya and the Netherlands case studies of the Ground Truth 2.0 project and participated in the team meetings of the two cases. These team meetings were held on a weekly or bi-weekly basis and aimed at planning and coordinating the establishment of the two CBM initiatives. The researcher's main role in these meetings was observer and despite limited and occasional engagement with the discussions in these meetings, the researcher did not have a decision making role in the establishment process of the two CBMs. The Ground Truth 2.0 team meetings are therefore considered as an additional source of observation for this research as they enhanced the researcher's understanding of the establishment process of the two CBMs.

The results and findings from the primary sources were further complemented by data from secondary sources such as review of relevant scientific publications, major laws, regulations, acts, government reports and statistics related to the issues in focus of the two CBM initiatives, as well as a number of relevant project documentations and reports (deliverables).

Finally, analysis of the tools and platforms developed in the two CBMs provided the researcher with complementary information for the case-specific analysis that is presented in chapters 4 and 5 of this dissertation.

3.3 DATA COLLECTION THROUGH INTERVIEWS - PHASE 1

3.3.1 Sampling methods and selection of interviewees

For collecting the data in first phase, i.e. the baseline data about the participation paradigms, power dynamics, and technological settings before establishment of the CBM initiatives, 60 in-depth semi-structured interviews were conducted in the Netherlands (26 interviews) and in Kenya (34 interviews) case. In general, four categories of stakeholders were approached for the interviews, namely participants in the co-design meetings of the two CBM initiatives, representatives of regulatory entities, members of the general public, and expert advisors (i.e. experts on the issue in focus of the CBM initiative).

The sample from the general public was selected using snowball and stratified sampling (based on both gender and age groups[6]). This category of interviewees included members of the general public who resided at the case study location, but were not involved in the CBM activities and did not know about the initiatives.

Interviewees from the other three categories of stakeholders were selected using cluster sampling. In case of the regulatory entities, at least one interviewee from the local or national authorities who has a formal or legal mandate for managing the issue in focus of the CBM initiative was interviewed. Expert advisors were selected from the pool of individuals with knowledge or experience about the issue in focus of the CBM initiative who either were identified in the co-design meetings or were already known to the researcher as part of his professional network. As an extra sampling criterion, at least 60% of the participants in the initial co-design meetings in the two cases were interviewed.

In many cases, the organizations and individuals that are categorized as regulatory entities or expert advisors were also present in the respective CBM co-design meetings and thus are also considered as members of the CBM co-design group.

Figure 3.1 illustrates the detailed distribution of interviewees in the two cases based on their age, gender and stakeholder type. It is important to note that the age and gender distribution of the samples are influenced by factors such as the overall population composition of each case and the (age and gender) composition of the participants in the CBM co-design meetings.

---

[6] The age groups are: (1) Under 18 years, (2) 18 to 25 years, (3) 26 to 35 years, (4) 36 to 45 years, (5) 46 to 55 years, (6) 56 to 65 years, and (7) 66 years or older

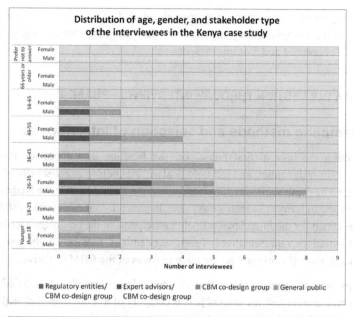

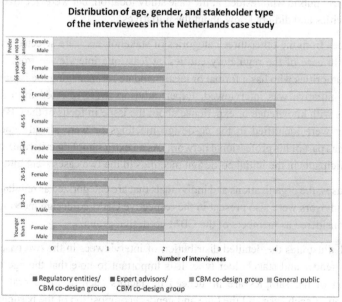

Figure 3.1 Age, gender and stakeholder type of interviewees in phase I in the two case studies

3.3.2 Design of interview protocols

The interviews were conducted using semi-structured interview protocols, which were designed based on the context-related dimensions and aspects of the CPI Framework. These interview protocols were designed using Google Forms. The complete lists of questions used for conducting the baseline interviews in the two cases are provided in Annexes 2 and 3. Because of the complexity of the topics and due to the fact that different stakeholders had different levels of knowledge or understanding about the issue in focus of the CBM, the researcher needed to customize the interview questions for the four categories of interviewees (see Annexes 2 and 3 for further details).

All interviews in the Kenya case study and some of the interviews in the Dutch case study were conducted in English; however, some of the interviewees in the Dutch case study were not comfortable with participating in an interview in English. The interview protocols were therefore translated into Dutch and one of the local partners of the Ground Truth 2.0 project (HydroLogic Research) conducted these interviews.

As a part of the design of the interview protocols, introductions, instructions and prompts were provided. Introductions were designed to introduce the topic of a group of questions that were asked in each part of the interview. Instructions were provided for the interviewers (other than the doctoral researcher) to guide them during the interview process. Moreover, where needed, prompts were provided to help the interviewees provide relevant answers for the questions.

3.3.3 Implementation of interviews and documentation of results

The interviews in this phase were conducted in the period between March 2017 and September 2017. The average length of the interviews was 50 minutes. Prior to conducting the interviews, all interviewees were presented with an informed consent form that included a brief introduction about the Ground Truth 2.0 project and the case study and clarified how their personal data and the information they provided would be used, stored and reported.

All interviews in the Dutch case study were conducted face-to-face. Because of the aforementioned language barrier in this case study, the majority of the interviews were conducted by a project partner (HydroLogic Research). In order to ensure the quality of the results, the PhD researcher designed a guideline document on how to conduct the interviews and held an instruction session to guide the two assistants prior to conducting the interviewers in this case study. After the interviews, the responses were translated back to English by the interviewers and they were stored in the designated Google Forms for this case study.

All interviews in the Kenya case study were conducted by the researcher. Five interviews were conducted during a field visit in May 2017, while the rest of the interviews in the

Kenyan case study were conducted via phone/Skype. Introductory calls were made to set up an interview appointments with the interviews and depending on the interviewee's access to internet, the interviews were conducted either over Skype or during a phone call. Similar to the Dutch case study, after the interviews, the responses were digitized and stored in the designated Google Forms for this case.

All interviews in this phase were coded with a three-part code for privacy purposes. If needed, these codes were used for referring to specific interviews, while reporting the results. Part one of this code is an abbreviation of the case study location, part two indicated the phase in which the data was collected (i.e. the first phase), and the third is a two figure number that refers to a specific interview, which is known to the researcher. For example, the code KE-01-06 refers to interview number 6, in the first phase of the interviews in the Kenyan case study.

3.4 DATA COLLECTION THROUGH INTERVIEWS - PHASE 2

3.4.1 Sampling methods and selection of interviewees

The second phase of the interviews aimed at understanding the process of establishment, functioning and results of the two CBM initiatives. Therefore, two main groups of potential interviewees who had enough knowledge about these processes and results were approached.

The first group included local stakeholders who participated in establishment of the CBMs or were among the end-users of the tools and processes developed in the CBM. This included representatives of community members and different organizations that participated in the co-design meetings, as well as end users and volunteers who participated in the data collection campaigns or trainings for using the CBM tools. Throughout this dissertation, this group is referred to as the *'CBM members'*. The second group consisted of the members of the Ground Truth 2.0 team who were involved in establishing the two CBMs, and therefore knew about the establishment process and results of each CBM. This group is referred to as *'GT2.0 team members'* hereafter. Potential interviewees from both groups were known to the researcher and members of the Ground Truth 2.0 team in each case; therefore, the sampling method in phase two of this research was cluster sampling. For selecting the sample from the first group, the researcher made sure to include at least one representative from the main organization who participated in co-design meetings. The sample from the second group included the Ground Truth 2.0 case study lead, as well as active team members from the main partner organizations who were involved in establishing the Dutch and the Kenyan case studies of the Ground Truth 2.0 project.

The sampling in this phase resulted in 32 in-depth interviews in the Dutch case study (12 interviews) and in Kenya case study (20 interviews). The sample in each case includes three members of the Ground Truth 2.0 team; the remainder of interviewees was from the CBM members. Figure 3.2 illustrates the detailed distribution of interviewees in the two cases based on their age, gender and interviewee categories. Similar to the interviews in phase 1, age and gender distribution of the sample population in each case is determined by the composition of the two aforementioned groups of potential interviewees.

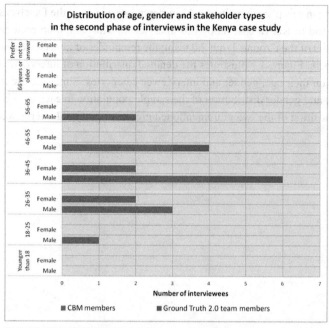

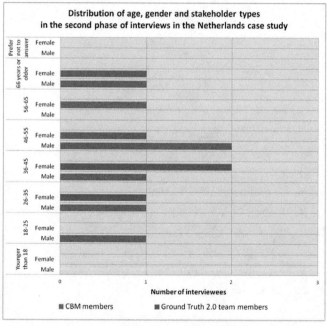

Figure 3.2 Age, gender and stakeholder type of interviewees in phase II in the two case studies

3.4.2 Design of interview protocols

Interviews in this phase were also conducted using semi-structured interview protocols; however, the interview protocols designed for this phase were based on all aspects and dimensions of the CPI Framework. The interview protocols were designed using Google Forms. Similar to the first phase, the two groups of interviewees had slightly different levels of knowledge or understanding about the process of establishment and functioning of the two CBMs. Therefore, two interview protocols, with slightly different questions were designed and used for conducting interviews with the two categories of interviewees. The complete lists of questions used for conducting interviews in this phase for the two cases are provided in Annexes 4 and 5.

In this phase, all interviews in the Kenya case study were conducted face-to-face and in English. Except for one interview with a former Ground Truth 2.0 team member who was interviewed over Skype, all other interviews in the Dutch case were also conducted face-to-face. However, only the interviews with the GT2.0 team members were conducted in English and all interviews with the CBM members were conducted in Dutch. The interview protocol for this group was therefore translated into Dutch.

As part of the design of the interview protocols, introductions, instructions and prompts and were provided for interviewers of HydroLogic Research. As a part of these instructions, the topic or focus of a group of questions were introduced in each part of the interview. Where needed, prompts were provided to help the interviewees trigger relevant answers to the questions.

3.4.3 Conducting the interviews and documentation of results

The interviews in this phase were conducted in October and November 2019. The average length of the interviews was 60 minutes. Similar to the first phase, prior to conducting the interviews, all interviewees were presented with an informed consent form that included a brief introduction about the Ground Truth 2.0 project and the case study and clarified how their personal data and the information they provide would be used, stored and reported.

Because of the aforementioned language barrier in this case study, only the interviews with the Ground Truth team members were held in English and all interviews with the CBM members were conducted by HydroLogic Research. In order to ensure the quality of the results, the PhD researcher designed a guideline document on how to conduct the interviews and held an instruction session to guide the two assistants prior to conducting the interviewers in this case study. After the interviews, the responses were translated back to English by the interviewers and they were stored in the designated Google Forms for this case study.

All interviews in the Kenya case study were conducted by the researcher. From the total number of 20 interviews in this case, all three interviews with the GT2.0 team members were held in Delft in October 2019. Three of the interviews with the CBM members were also held in Delft in October 2019, while they were visiting IHE Delft for a Ground Truth 2.0 project event (i.e. Ground Truth Week 2019). The remaining 14 interviews were conducted during a field visit in November 2019. Similar to the Dutch case study, after the interviews, the responses were digitized and stored in the designated Google Forms for this case.

Similar to interviews in the first phase, all interviews in this phase were also coded with a three-part code for privacy purposes. For example, the code NL-02-03 refers to the third interview in the second phase of the interviews in the Dutch case study.

3.5 DATA ANALYSIS METHODS

As explained earlier, the results of this research are based on primary empirical data from the interviews and observation protocols, secondary data such as publications, reports and project documentations, and also analysis of the developed tools and platforms in the two CBMs. This section provides an overview of methods used by the researcher for analyzing this data.

Due to the complexity of the research topic and also the large number of interviews in the two phases of this study (i.e. 92 interviews in total), the researcher decided to use a qualitative data analysis software called MAXQDA for analyzing the interview responses. MAXQDA allows for storing, coding, categorizing, and comparing qualitative data in a systematic and structured way. A separate MAXQDA model was developed for each case study in each phase of the research. Figure 3.3 provides an example of a MAXQDA model that was developed for the analysis of the interview results. The process of analyzing the interviews in each phase and for each case study started with exporting the responses from the Google Forms to excel, and these excel files were then imported to MAXQDA. Each individual interview was quickly scanned and a first impression from each interview was noted. Next, each interview question was analyzed across the sample (or within a specific group of interviewees in the sample). Different segments of the responses were coded using the code function in MAXQDA. Created codes were then categorized, merged or renamed in the course of analysis. The 'memo' function in MAXQDA was used for recording the findings for each code and question. The analysis window of the software facilitated making different queries across the results of the interviews.

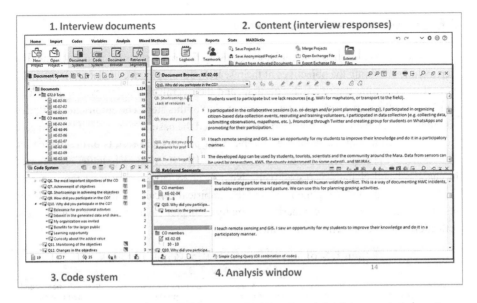

Figure 3.3 Example of the MAXQDA model developed for the second phase of interviews in the Kenyan case study

According to the main purpose of this research, i.e. evaluating the factors that influence the establishment, functioning and outcomes of CBM initiatives, the unit of analysis is the CBM initiatives in each case study. Nevertheless, the information from the interviews reflect perception of individuals (or individuals within organizations), and thus the unit of observation in this research is individuals.

In order to complement the findings from the interviews, and also for triangulation purposes, observations and personal notes (during stakeholder meetings, team meetings and field visits), GT2.0 logbooks, relevant publications, reports and project documentations were analyzed using a deductive approach. The results of this analysis are reported and referred to throughout this dissertation, especially in chapters four and five.

Furthermore, as an additional source of information, the tools developed in each case study were analyzed and the results of this analysis were integrated into the results and discussions of this research. This includes for example a review of the contents of the web-platforms of each CBM initiative or trend analysis of the data submitted in the cases.

3.6 SUMMARY OF RESEARCH DESIGN

This research was designed and implemented using five main steps (Figure 3.4). The first step included defining the objectives of the research, developing a conceptual framework, as well as designing the research questions and the overall methodology. Chapters 1, 2 and 3 of this dissertation present the documentation of this step. The empirical qualitative research was conducted in steps 2 and 3 and included the baseline analysis of the two case studies, as well as evaluation of the establishment process and results of the two cases. The results of the empirical research in the Dutch and the Kenyan case studies are reported in chapter 4 and chapter 5, respectively. Next, a cross cases analysis of the two case studies was conducted using the results of the empirical research. Chapter 6 presents the results of this cross-case comparison. Finally, Chapter 7 presents the conclusions, reflections and recommendations of this study.

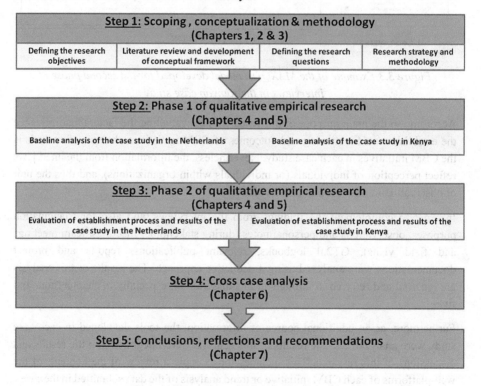

Figure 3.4 Summary of the research design

3.7 Conclusion

This chapter presented the methodological choices and steps that were taken for conducting this research. The topic of this research was described as an example of a complex and contemporary phenomenon that needs to be studied using an enabling methodology. Therefore, a two phase qualitative empirical research and a case study approach were chosen as the main methodology. This chapter also introduced the sources of data for this research. Due to the fact that interviews are the main source of empirical data for this study, the steps involved in designing the interview protocols, selecting the samples and implementing the interviews in the two phases of research were explained in detail. Moreover, data analysis methods were detailed and the use of MAXQDA for analyzing the empirical data in this research was explained. Overall in the two phases of empirical research 92 in-depth semi-structured interviews were conducted across the two case studies and the results of these interviews informed the findings and discussions about the baseline situation, establishment process, and results of the two cases. Chapter 4 and 5 of this dissertation are dedicated to presenting the findings in the two case studies.

4

RESULTS AND DISCUSSION OF THE DUTCH CASE STUDY: GRIP OP WATER ALTENA[7]

The main aim of this chapter is to present the results of applying the conceptual framework of the research (i.e. the CPI Framework) for evaluating the baseline situation of the Dutch case study, as well as the establishment process and results of the CBM in this case, i.e. Grip op Water Altena. This is in line with the second and third objective of this research that is testing the empirical applicability of the conceptual framework and evaluating the evolving processes, outputs and interim outcomes of the CBMs over time. Section 4.1 provides background information about the Dutch case study and aims to familiarize the reader with this case and Grip op Water Altena. Section 4.2 presents the results of the baseline analysis in the Grip op Water Altena. The content of this section is informed by the first phase of qualitative empirical research. Section 4.3 describes the establishment process and the results of Grip op Water Altena. Presented results in sections 4.3 are informed by the second phase of qualitative empirical research of this study. Finally, the findings of this case study are discussed in sections 4.4 and 4.5.

---

[7] This chapter is partially based on: Gharesifard, M., Wehn, U., & van der Zaag, P. (2019a). Context matters: a baseline analysis of contextual realities for two community-based monitoring initiatives of water and environment in Europe and Africa. *Journal of Hydrology*, 124144. doi:https://doi.org/10.1016/j.jhydrol.2019.124144

4.1 BACKGROUND OF THE DUTCH CASE STUDY

The Dutch case study is located in the 'Land van Heudsen en Altena', which is a part of the Dutch province of North-Brabant. In terms of water management, this area falls under the jurisdiction of the Regional Water Authority Rivierenland. At the start of this study the area consisted of the three municipalities of Werkendam, Woudrichem and Aalburg. On 1st of January 2019, these municipalities were merged and formed the new municipality of 'Altena' (Figure 4.1). This new municipality has a total surface area of 211 km^2 and a population of 55,840 inhabitants[8].

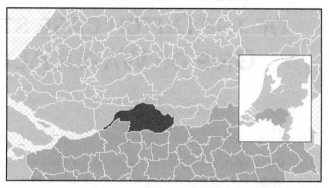

Figure 4.1 Location of Altena in the province of North-Brabant and the Netherlands Source: Wikipedia[9]

In the past, Land van Heudsen en Altena used to get flooded regularly from the rivers surrounding it; a problem that triggered the damming of a part of the rivers (Vergouwe, 2016), and a number of other structural measure in this areas. In recent years, the flooding from the rivers has not been a problem; however, with an increase in the number of intense rainfall events in recent years, pluvial flooding has become a major concern and has negatively affected a number of residents in this area. For example on July 28, 2014 and 30 to 31 August 2015, residents of Altena witnessed two extreme rainfall events, during which parts of this area received more than 125 mm of rain in just a few hours (Figure 4.2).

---

[8]   Source of the surface area and population figures: CBS Statline (in Dutch). Retrieved from https://opendata.cbs.nl/statline/#/CBS/nl/?fromstatweb, 05 November 2019.
[9] By Michiel1972, derivative by Thayts - Derived from File:Map - NL - Municipality code 0392 (2019).svg and File:Map - NL - Municipality code 0870 (2009).svg, CC BY-SA 3.0, https://commons.wikimedia.org/w/index.php?curid=83010132

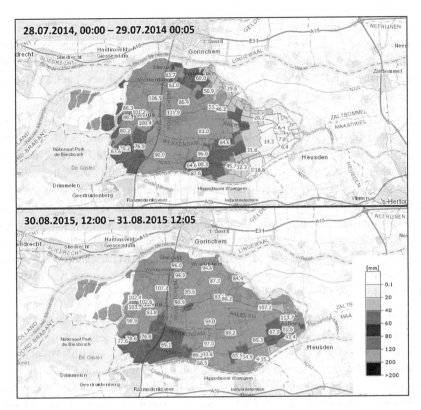

Figure 4.2 Extreme rainfall events of 2014 and 2015 in Altena
Source: http://altena.gripopwater.nl/

In 2016, and as a part of the Ground Truth 2.0 project, a CBM was planned to be established in Land van Heudsen en Altena, using a co-design process. According to the proposal of the Ground Truth 2.0 project, the initial focus of this CBM was on the issue of pluvial flooding (Ground Truth 2.0 consortium, 2015). This initial focus was first identified at the proposal stage of the Ground Truth 2.0 project and by HydroLogic Research (i.e. one of the project partners). At that time the Regional Water Authority Rivierenland was a client of HydroLogic Research and they knew that this water authority has an interest to work more closely with citizens in Altena. The initial ambition of this CBM was therefore to facilitate communication and information exchange on water availability between the Regional Water Authority Rivierenland and local citizens (Ground Truth 2.0 consortium, 2015). It was expected that such a CBM would contribute to better policy and decision making about management of pluvial flooding and in turn contribute to creating a more water-resilient Altena.

HydroLogic Research and IHE Delft Institute for Water Education were the core partner organizations that facilitated the co-design process of the CBM in the Dutch case study of the Ground Truth 2.0 project. The Ground Truth 2.0 team members were therefore from these two organizations. The number and composition the team members fluctuated over time and was determined by factors such as the amount of support needed and staff turnover in the aforementioned organizations. Nevertheless, a group of at least five team members were constantly involved in the establishment process of this CBM.

In 2016, Ground Truth 2.0 team members approached the Regional Water Authority Rivierenland and invited this organization to become a member of this CBM initiative. The water authority showed an interest in the topic of the project because they were curious about the added value of the initiative for their activities, especially monitoring of, and awareness raising about, pluvial flooding. Therefore, they helped with reaching out to other stakeholders, including a number of active local community members that they knew from previous projects in the area and the three aforementioned municipalities. The first co-design meeting of this CBM was organized in May 2017, with 21 participants including representatives of the Regional Water Authority Rivierenland, municipalities of Werkendam, Woudrichem and Aalburg, a few environmental NGOs, as well as a number of local community members.

The aims, objectives and functionalities of this CBM were co-created during 5 co-design meetings between May 2017 and January 2018. The first two co-design meetings were dedicated to agreeing on the vision, mission and objectives of this CBM, while the next three sessions focused on designing the functionalities and tools in this CBM. The agreed upon goal of stakeholders in this CBM was to prevent damage from extreme precipitation and this is reflected in the name of the initiative, 'Grip op water Altena' (in Dutch), which means grip on water. This was planned to be achieved by improving communication and data, information and knowledge sharing between local citizens, researchers, three aforementioned municipalities and the Regional Water Authority Rivierenland.

Based on the outputs of the co-design meetings the Grip op Water Altena web-platform[10] was designed that was launched in November 2017. This platform can be used by local stakeholders to raise awareness and communicate about the issue of pluvial flooding in Land van Heudsen en Altena. Inventorizing measures taken by authorities and citizens for reducing problems with pluvial flooding, reporting problems with pluvial flooding, generating and sharing tips and tricks on how citizens can contribute to reducing problems with pluvial flooding and generating a better overview of available water storages in the area (using a garden survey) are examples of information exchanged via this web-platform. The content of this web-platform evolved and additional information became available as the CBM progressed with its activities. Annex 6 presents a number of screenshots from the Grip op Water Altena Web-platform.

---

[10] http://altena.gripopwater.nl/

Becoming a member of Grip op Water Altena was free and anyone who knew about the CBM and wished to participate could attend the meetings or join online activities. After the co-design phase, CBM members participated in 9 subsequent face-to-face planning meetings to discuss activities and future sustainability of this CBM. Moreover, CBM members used a number of local public events to introduce their initiative to the wider public and recruit more volunteers. Boerenerfdag in August 2018 and Molendag in May 2019 are examples of these public outreach events.

Regardless of these efforts, the number of participants in the face-to-face meetings decreased over time and not many users regularly used the Grip op Water Altena web-platform. Moreover, alignment with well-established and quite structured ways of working in the water authority and municipalities proved to be difficult and this CBM could only partly achieve its ambitions. The factors that affected the establishment process, functioning and results of Grip op Water Altena were both internal to this imitative and context-related. The context in which Grip op Water Altena was being established is explained in section 4.2 and the process of establishment and results of this CBM are discussed in detail in section 4.3.

4.2 BASELINE SITUATION OF THE DUTCH CASE STUDY

This section is dedicated to presenting the findings of the baseline analysis of the Dutch case study that is informed by the phase 1 of the empirical research. In order to be able to capture the baseline situation, in which Grip op Water Altena was being established, the elements of the CPI Framework that focus on the contextual aspects were studied. Specifically, these elements include the context-related aspects of 'power dynamics', 'participation' and 'technology' dimensions (Figure 4.3).

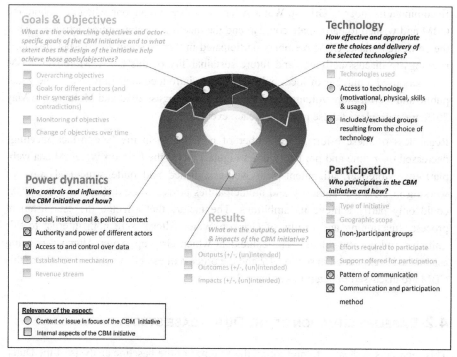

Figure 4.3 Framework for examining the contextual settings of CBM initiatives
Source: Gharesifard et al. (2019a)

4.2.1 Power dynamics in the Dutch case study

Social, institutional & political context

There are three major institutional layers in the Dutch water management system; (1) European level, including the EU and International River Basin Commissions; (2) national level, which includes the central government (Ministry of Infrastructure and Water Management) and the national Water Authority (Rijkswaterstaat); and (3) sub-national level, which consists of the 'triangle' of provinces, Regional Water Authorities and municipalities (OECD, 2014). Dutch water management has been traditionally decentralized from both, functional and territorial perspectives (OECD, 2014). Functional decentralization refers to the tasks that each decentralized unit carries, while territorial decentralization is linked to different provinces and municipalities within the country. In terms of hierarchy, the Regional Water Authorities and municipalities are both at the same level, while the provinces have a supervisory function over both. The provinces are also responsible for laying down rules about water management and setting up or dissolving Regional Water Authorities (Havekes et al., 2017).

During the past decade, the issue of pluvial flooding has moved from being absent in the Dutch water policy for flood protection, climate adaptation and water resilience to being high on the political agenda since 2016 (Delta Programmes 2016, 2017 and 2018). In practice, the main actors involved in managing pluvial flooding are the municipalities and Regional Water Authorities. This is due to the fact that the Water Act (2010) places the municipalities in charge of processing 'urban waste water', which includes rainwater run-off, in the municipal area and obliges the municipalities and Regional Water Authorities to coordinate their activities. The issue of pluvial flooding in urban areas is closely linked to spatial planning. Municipalities and property developers are the main stakeholders who benefit from spatial development in the Netherlands; nevertheless, unlike the Regional Water Authorities, they do not bear much of the cost when it comes to water management (OECD, 2014), nor is water management among their primary tasks. Moreover, the Regional Water Authorities are independent from the central government in terms of generating their own revenue, while municipalities are largely dependent on the national government for their income via municipal funds.

In addition to the Water Act (2010) and the Delta Programmes, there are several other laws, agreements, plans and policy guidelines related to the topic of this CBM initiative, among which are the National Administrative Agreement on Water, the Spatial Planning Act, the National Water Plan 2016–2021, the Noord-Brabant Provincial Environment and Water Plan 2016–2021, the water management plan of the Rivierenland Regional Water Authority, the sewerage policy vision of the Werkendam, Woudrichem and Aalburg municipalities and the common water plan of the municipalities and the Rivierenland Regional Water Authority. Moreover, at the time of conducting this research, the Dutch government was working on developing the 'Environmental Planning Act' (expected to take effect in 2021); an Act that will combine several existing laws, including the Water Act, the Crisis and Recovery Act and the Spatial Planning Act into one simplified and coherent piece of legislation.

Almost all interviewees believed that the implementation of the rules, roles and responsibilities related to the management of pluvial floods is very strict and in some cases even stricter than the available protocols. Several interviewees considered the fact that they do not see much of a problem with pluvial floods in the Netherlands as a sign of good implementation. Moreover, a number of interviewees mentioned that they are satisfied with the fact that there are numerous agreements, protocols and procedures in place and being implemented, especially in emergency situations.

The majority of the interviewees argued that the authorities are responsible for managing issues such as pluvial flooding and that civilians pay for this service via taxes. The interviewees also believed that trust in the government is deeply rooted in Dutch culture and that Dutch citizens do not want to think about such problems too much. Several interviewees from all stakeholder groups mentioned the Dutch quickly shift attention to matters visible for them. For example, in case of an incident (e.g. a major flood), the Dutch community tends to come together and help voluntarily to prevent loss of lives or damage to properties. The result of the interviews also revealed an 'awareness gap' among the Dutch citizens regarding water management, an issue that was also highlighted in OECD (2014). This awareness gap, combined with high level of trust in the government, can hinder the implementation of necessary actions in the prevention phase of an infrequently experienced problem such as pluvial flooding.

Baseline situation of authority and power of stakeholders

All interviewees from regulatory entities in this case mentioned that they exert influence on decisions related to pluvial flood management by providing advice/consultation, joining other colleagues or other organizations for decision making, or by having direct authority on the decisions because of their position. Expert advisors who are not part of regulatory entities mainly have an influence by providing advice to the authorities, or communicative influence, for example by publishing their research findings.

The majority of the interviewees from the CBM co-design group and the general public perceived to have little or no influence on decisions related to managing pluvial floods, or mentioned that they have an indirect influence via participation in the elections of Regional Water Authorities. Several participants also mentioned they do not (automatically) receive information about such decisions; however, if they are interested, they can search for and find this information. On the other hand, a few interviewees from the CBM co-design group and the general public gave examples of communicative influence on decisions, such as signing a petition, writing a letter to the authorities and expressing their opinions on how to manage pluvial floods, or campaigning via social and mass media. Regarding the latter, one of the interviewees argued that "you can always complain by publishing an article in a newspaper or writing a note on their Facebook page; 'public shaming' works"[11].

Baseline situation of access to and control over data and information

Data and information in the context of the Dutch case relates to both data about the environmental issue in focus and information about different steps of policy and decision making processes. This includes for example data about rainfall and water levels, as well as technical and financial information about projects that aim at reducing the damage from pluvial flooding. In line with Article 110 of the Constitution of the Kingdom of the Netherlands, all Dutch government information has to be shared with the public and any information that is not already publicly shared (e.g. via websites) can be requested in

---

[11] NL-01-18

accordance with the Dutch Public Access to Government Information Act (Law of 31 October 1991). This law, however, also allows government organizations to withhold certain information if considered confidential for privacy or security reasons. Public authorities have explicit deadlines for responding to requests, and in case of refusal should provide the applicants with the reasons for their decision.

The interviewees from the regulatory entities and the expert advisors believed that a lot of data is available about pluvial floods, but the general public is not aware of this because the information is mostly stored in internal databases within organizations and thus not immediately available for the public. The majority of the remaining interviewees, especially those from the general public, did not have any experience with accessing data about pluvial floods; however, the few who did, mentioned that some organizations, e.g. the Dutch Meteorological Institute (KNMI), make more data publicly available than others.

The interviewees identified a wide range of organizations and individuals as qualified to analyze data about pluvial floods, including the Ministry of Infrastructure and Water Management, Rijkswaterstaat, the Regional Water Authorities, the Union of Regional Water Authorities, provinces, municipalities, KNMI, NGOs, consultancies, scientists and scientific institutes. Only a few interviewees considered citizens among this group.

The Regional Water Authorities were identified as the main organization that collects data about pluvial floods at the local level and Rijkswaterstaat was identified as the main producer and owner of data at the national level. In addition, some interviewees believed that municipalities and provinces hardly take measurements themselves, and even if they do, the measurements are mostly qualitative (e.g. whether or not a sewage overflow occurred), not quantitative (e.g. how much sewage water has been released to the surface waters). Regarding control over data, it was mentioned that it is mainly the owner of the data who defines the level of access for others.

4.2.2 Participation dynamics in the Dutch case study

Baseline situation of participation in policy and decision making processes

The interviewees were asked to identify those stakeholders who they think are currently involved in decision making and policy making processes regarding the management of pluvial floods in the Netherlands. The identified stakeholders included the major actors involved in water management in the Netherlands, namely the Regional Water Authorities, municipalities, Provinces, Rijkswaterstaat, Ministry of Infrastructure and Water Management, Union of Regional Water Authorities (Unie van Waterschappen), the Association of Dutch municipalities (Vereniging van Nederlandse Gemeenten). Moreover, the interviewees also identified the Rioned Foundation, an umbrella organization for urban water management and sewerage in the Netherlands, and the Southern Agriculture and Horticulture Organization (ZLTO), which is a farmers association with 15,000 members in Zeeland, Noord-Brabant and Zuid-Gelderland. In addition, the interviewees also perceived an advisory role for research and consulting companies from the private sector in managing pluvial floods.

The role of the citizens, however, was mentioned to be limited to electing representatives in the parliamentary, provincial, municipal, or Regional Water Authority elections. This is due to the fact that public participation is not a fundamental pillar of Dutch democracy. There is no mention whatsoever of public participation in the Constitution of the Kingdom of the Netherlands, nor in the Water Act (2010). Although citizen participation is gaining more attention in more recent policy guidelines such as the National Water Plan 2016–2021, fundamentally, it is "at best, as an instrument to improve the current working of representative democracy" (Michels, 2006, p. 336).

Existings patterns of communication

The Safety Regions Act (2010) sets the legal framework for risk and crisis communication in the Netherlands. This Act divides the country into 25 security regions, which are extended local government units with joint safety regulations. The management board of a security region is responsible for informing a number of government authorities and personnel as well as the citizens about risks and crises that may affect a specific region. Although occasional bi-directional and interactive patterns of information flow for risk and disaster communication can be found within and between (central) government authorities, the predominant pattern of information flow is unidirectional, from authorities to the citizens. In addition, this Act only focuses on such communication at regional level; however, as Kaufmann et al. (2016) clarified, communication of preparatory and responsive measures for events such as pluvial floods at a smaller scale can highly differ from one location to another.

The interviewees were asked whether or not they communicate about pluvial flooding with others and to indicate their preferred channels for such communication. Using Apps on smart phones (e.g. WhatsApp, Viber, Line) and communicating via email, telephone (call or SMS), or face-to-face (e.g. daily bilateral discussions with others or participation in meetings) were among the most frequent means of communication in this case study. Interviewees from the regulatory entities and expert advisors indicated that they communicate about this issue mostly for work purposes, both internally (e.g. with colleagues) and externally (with other individuals and organizations, e.g. with residents or Regional Water Authorities). Interviewees in this case often linked the issue of pluvial flooding with bad weather and intense rainfall events. The interviewees from the CBM co-design group and the general public mentioned they often communicate about the issue with other residents in informal settings or after heavy rainfall. At the same time, a number of interviewees from these groups emphasized that unless there is an actual problem or an emergency situation, there is no or little communication with municipalities and other authorities. Figure 4.4 summarizes the preferred channels for communicating. Based on this figure, the most frequently mentioned preferred channel for communication was phone call or SMS, as people prefer to receive a quick response or like someone listen to them.

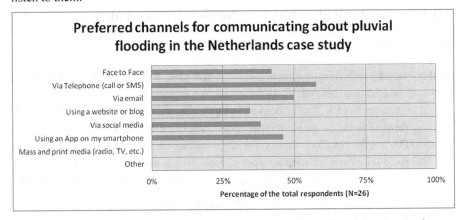

Figure 4.4 Preferred channels for communicating about pluvial flooding in the Netherlands case study
Source: Gharesifard et al. (2019a)

Existing methods of communication and participation in decision making processes

The interviewees were asked to explain in what ways (if any) they are involved in managing pluvial floods and how they take part in decision making processes on this issue. All interviewees from the regulatory entities mentioned that they are involved in managing pluvial floods via their jobs. Their involvement spanned from setting policies at the Regional Water Authority or municipal level to decision making at the operational level (e.g. managing the pumps and weirs for preventing floods). The examples provided by the interviewees from this group depicted the 'deliberate and negotiate' or 'technical expertise' roles for them in the decision making processes (Fung, 2006).

The expert advisors are mostly involved indirectly via providing data and information, as well as expressing their expert opinion in meetings with the authorities. These data, information and expert views may or may not be used by the authorities. Thus they are mostly involved in decision making processes via explicit data provision and expressing preferences.

The majority of the interviewees from the CBM co-design group and the general public had little or no expectation to influence decisions at any level. In this regard, one of the interviewees mentioned that "I don't think I have any influence, it is the municipality who

61

makes the decisions. If I have a problem, I contact the municipality and they decide how to take care of the issue"[12]. A few interviewees, however, believed that they have an indirect say in decisions by participating in the elections of the Regional Water Authority elections, but they did not consider this as 'real' involvement in making decisions or creating policies.

4.2.3 Technological context in the Dutch case study

Baseline situation of access to technology

At the time of conducting this study, with an IDI rank 7 (ITU, 2017a) and a Digital Economy and Society Index (DESI) 4 (European Commission, 2017), the Netherlands ranked high in terms of ICT development in Europe as well as worldwide. Fulfilling the growing demand for high-speed Internet access is at the forefront of Dutch government policy (ITU, 2017b). In 2017, there were 129.9 mobile-cellular telephone subscriptions and 40.3 fixed-telephone subscriptions per 100 inhabitants in the Netherlands. Moreover, 92.2% of households had internet access and almost 90% of households had computers.

With 90.4% of citizens using the internet, the country also ranked high in terms of actual use of information and communication technologies. Dutch citizens also have the required skills to use ICTs. According to the ITU (2017a), the 'mean years of schooling' in the Netherlands is 11.9 years. Moreover, the Dutch government aims to make digital literacy a core part of the curriculum of primary and secondary education (Ministry of Economic Affairs, 2016).

Baseline situation of included/excluded groups from use of technology

Despite a well-developed and growing ICT infrastructure and availability of affordable mobile-cellular and mobile-broadband offers (ITU, 2017b), at the time of conducting this research, the access gap was not yet closed in the Netherlands. The results of a recent study of 108,000 Dutch citizens reveal that the younger population, with higher levels of education and income and (in some areas) males have better access to internet (van Deursen & van Dijk, 2015). In terms of actual usage, according to available statistics, a very large proportion of the Dutch population is active technology users. Nevertheless, the main gap resulting from the growing digitization of the Dutch society seems to exist for those who lack basic digital skills. According to the European Commission (2017), this is no less than 23% of the Dutch population aged 16–74; a thought provoking figure but one that is still much lower than the 44% average for the whole Europe.

---

[12] NL-01-14

4.3 THE ESTABLISHMENT PROCESS AND RESULTS OF GRIP OP WATER ALTENA

This section describes the establishment process and results of Grip op Water Altena, based on the findings of the second phase of empirical research in this case. Sub-sections 4.3.1 to 4.3.5 correspond with the five dimensions of the CPI Framework and help answer research questions 1 to 5 of this study. It is important to mention that the results reported in this section reflect the status of Grip op Water Altena at the end of November 2019 (i.e. the end date of data collection in phase 2 of this research).

4.3.1 Objectives and actor specific goals in Grip op Water Altena

Based on an evaluation of the overarching objectives and actor-specific goals in Grip op Water Altena, this section provides the answer to the first research question for this CBM, i.e. what are the overarching objectives and actor-specific goals of this CBM and to what extent does the design of this initiative help achieve those goals/objectives?

Overarching objectives of Grip op Water Altena

The co-design process of Grip op Water Altena, resulted in defining a common vision and mission, as well as five specific objectives for this CBM, which are presented in Table 4.1.

Table 4.1 Vision, Mission and Objectives of Grip op Water Altena

Vision
"In Land van Heusden en Altena the municipalities, water authority, citizens and farmers understand each other's interests and ways of working and are together responsible for limiting the damage by pluvial flooding in urban and rural areas"

Mission
"The citizen observatory is a place (on- and offline) where collected observations, knowledge and warnings are shared, where bottlenecks and measures are constructively discussed along short communication lines and where it is clear which actions are taken by which party"

Objectives
Obj1. "Facilitate the exchange of observations and information about the weather and water systems [in October 2017] to allow all stakeholders to act or plan ahead"
Obj2. "Support short communication lines and insight in plans and activities of stakeholders regarding water management in Land van Heusden en Altena [early 2018]"
Obj3. "Set up a knowledge platform with action perspectives and tips to take measures against damage from pluvial flooding [in the course of 2018]"
Obj4. "Support an open and constructive dialogue between all involved parties in Land van Heusden en Altena [from the start] and expand the network towards a real water community"
Obj5. "Prepare the sustainable continuation of this CBM after Ground Truth 2.0 [in 2018 and 2019]"

Interviewees from both the GT2.0 team and members of Grip op Water Altena were asked to reflect on the objectives of this CBM by (1) indicating their perception about the most important objective(s), (2) reflecting on the extent of achievement of the objectives, and (3) (if any) identifying shortcomings in achieving the objectives.

The GT2.0 team members believed that the third and the fourth objectives of Grip op Water Altena are the most important objectives of this initiative and mentioned that this CBM mainly focuses on sharing information, practical tips and raising awareness on how to prevent pluvial flooding in a collaborative and participatory way. Moreover, one of the interviewees mentioned that there seems to be a knowledge gap among the Dutch citizens about the measures taken by municipalities and Water Authorities. This interviewee indicated that Grip op Water Altena aimed at creating new opportunities for a two way communication between authorities and citizens, in line with its fourth objective.

Similarly, the majority of the CBM members in this case identified the third and the fourth objectives as the most important. Most interviewees described Grip op Water Altena as an initiative that aims at knowledge creation and awareness raising about the topic of pluvial flooding in Altena, and how this issue can be addressed in a participatory way. Awareness raising was both for the citizens, to realize the ways by which they can contribute to reducing the risk of pluvial flooding, and for the authorities to examine that this can be done in a collaborative way and with participation of all stakeholders. Moreover, Grip op Water was described as a platform that is being used by the stakeholders for communicating, sharing information, and learning from each other.

All interviewees believed that the objectives of Grip op Water Altena have been partly achieved. Based on the results of the interviews, the extent of achievement of the third and the fourth objectives of this CBM is far more than the first, second and fifth objectives.

The GT2.0 team members believed that this initiative was not particularly successful with engaging a bigger number of community members and this contributed to fewer possibilities for achieving its rather ambitious and broad objectives.

Limited engagement of community members was also the most frequently mentioned shortcoming by the CBM members. It was mentioned that the initiative started with a larger group; however, instead of growing in size, a number of people dropped out and left the initiative. The actual number of participants in the meetings confirms this, as the first co-design meetings took place with participation of more than 20 local stakeholders, while towards the end of the project, there were usually a group of less than 10 participants attending. With this regard, some interviewees believed that it took a very long time for Grip op Water to find its purpose and chose its direction in a collaborative way and this resulted in discontinued participation of community members and organizations. This was partly due to the fact that while establishing the CBMs in different cases, the Ground Truth 2.0 team members were also busy with developing the co-design methodology itself. Moreover, interviewees believed that people who left Grip op Water either did not see tangible results being produced in the initiative, or did not perceive the issue in focus of this CBM to be urgent. In this regard, a member of the Regional Water Authority Rivierenland mentioned, when there is a flood incident, a lot of people are enthusiastic to join. However, we have had no flooding for a while and therefore people forget or do not feel the urgency to join an initiative like Grip op Water anymore[13].

Goals of different actors in Grip op Water Altena

Overall, the water authority, municipality and community members (including individuals and organized community groups) formed the three main stakeholder groups in Grip op Water Altena, and each had their own goals and interests for participation in this initiative.

When asked about the reason for their participation, the representatives of the water authority mentioned that their organization was invited by the Ground Truth 2.0 project, because of their expertise in the area of water management. They joined the CBM because they were curious about its added value for their activities, especially monitoring of, and awareness raising about, pluvial flooding. Therefore, they helped with reaching out to other stakeholders.

Interviewees from the municipality mentioned that they joined the Grip op Water Altena meetings, because they wanted to create a better understanding between citizens, the municipality and other organizations about management of pluvial floods and also because they wanted to improve their water management system.

Interviewees from the local community mentioned three main reasons for their participation. The first one was interest in nature and their living environment. They found Grip op Water Altena an opportunity to be part of conversations about water management in Altena. The second reason was that they wanted to know more about how the municipality and water authority operate, and the measures taken by them to reduce the problem of pluvial flooding in Altena. Thirdly, some community members wanted to have a short communication line with the authorities in charge of managing pluvial floods. They found such a communication channel not only important for contacting the authorities in case of an emergency, but also for expressing concerns and sharing ideas.

Monitoring of the objectives in Grip op Water Altena

There was no formal procedure specifically designed for monitoring the objectives of Grip op Water Altena. Two of the three interviewees from the Ground Truth 2.0 team mentioned, given the fact that the objectives of this CBM were defined too broadly; they found it difficult to measure the achievement of the objectives. Nevertheless, the GT2.0

---

[13] NL-02-07

team members identified two mechanisms that helped with monitoring and reflecting back on achieving the Grip op Water Altena objectives.

The first mechanism was revisiting, and informally taking stock of, the objectives during collaborative meetings. From the moment that the objectives of Grip op Water Altena were co-created and agreed upon, these objectives were revisited regularly in almost all face-to-face meetings with the stakeholders. This provided the GT2.0 team members and all stakeholders with an opportunity to revisit and reflect back on the objectives.

The second mechanism was through a practice called 'reverse impact journey' (also known as reverse objectives journey); a method based on the logic model and theory of change (Friedman, 2008), which was used by the Ground Truth 2.0 team to reverse engineer the activities that are needed for achieving the agreed-upon objectives of Grip op Water Altena. The idea of reverse impact journey was first discussed among the GT2.0 partners in May 2018 and therefore until then this mechanism did not guide monitoring of the objectives in this or other CBMs of the Ground Truth 2.0 project. Starting from the objectives the Ground Truth 2.0 team mapped out the outputs and outcomes that would have been expected from achieving each objective and then identified the activities that needed to take place for producing those outputs and outcomes. For example, the third objective of Grip op Water Altena, which is setting up a knowledge platform with action perspectives and tips to take measures against damage from pluvial flooding, is only achievable through commitment of different actors for uploading tips and action perspectives to the CBM web-platform. This means that there was a need for identifying and assigning roles and responsibilities to different actors regarding data and information generation and exchange.

Change of the Grip op Water Altena objectives over time

All three interviewees from the GT2.0 team indicated that the objectives of Grip op Water Altena were (unofficially) changed during the course of its establishment. Particularly, the first and the second objectives got out of focus and more emphasis was put on achieving the third and the fourth objectives.

The first objective that focused on collecting and sharing data about weather and water systems was pushed to the sidelines because the stakeholders could not agree on the content and procedure of such observations. Several ideas for making observations were discussed during the collaborative meetings; however, these ideas were mostly turned down by the representatives of the municipality. Waterlabel[14], was an example of these proposed observation; an initiative that allows for scoring individual buildings based on their capacity for water retention. The idea behind Waterlabel is similar to the concept of

---

[14] https://www.waterlabel.net/

67

energy label for homes. The municipality of Altena tested the Waterlabel website, but was not in favor of promoting its use. This was because based on the results of their test, they came to the conclusion that the methodology behind Waterlabel will not allow for capturing improvements in the water storage capacity of the homes.

The second objective of Grip op Water Altena that aimed at supporting short communication lines between different stakeholders also gradually got out of focus. At the time of formulating the objectives of this CBM, the issue of short communication lines was so important that it found its way to the mission of Grip op Water Altena. However, except for the face-to-face meetings and a WhatsApp group (with eight local stakeholders), no other short communication lines was set up for discussions among the stakeholders. The main reason for not focusing on this objective was that the role of the officials involved in urban water management (i.e. the municipality and the water authority) is heavily institutionalized and therefore, there was little or no room for change in the way they were willing to communicate with other stakeholders. Disclaimers on the Grip op Water Altena, such as the one shown in Figure 4.5, depict the inflexibility of the government organizations for using alternative channels of communication. This note for citizen-contributed observations of pluvial flooding reads; "your observation does not automatically end up with the municipality or the water board. If you want to make a report, contact the municipality or the water board".

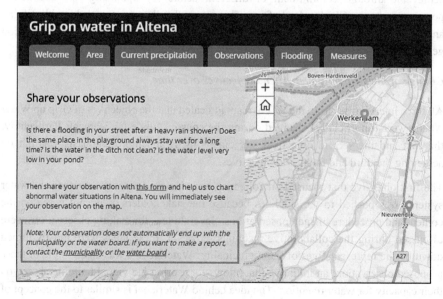

Figure 4.5 An example of a disclaimer on Grip op Water Website

4.3.2 Participation dynamics in Grip op Water Altena

This section aims at unpacking the participation dynamics in Grip op Water Altena, which help answering the second research question of this study; who participates in the CBM initiative and how, and who does not?

Type of initiative

During the functional design of the Ground Truth 2.0 project, based on the higher aim of the CBMs, a distinction was made between three types of initiatives. This categorization is referred to as the 'citizen observatory domains', and includes 'Environmental Monitoring', 'Cooperative Planning' and 'Environmental Stewardship' (Wehn et al., 2015a). In this conceptualization, Environmental Monitoring refers to CBM initiatives that focus on implicit and explicit data collection and sharing by the members of the public. Cooperative Planning domain includes initiatives that form interactive activities such as consultation, discussion and feedback among different stakeholders. Finally, 'Environmental Stewardship' refers to CBM initiatives that focus on creating shared responsibilities and collaboration between different stakeholders for addressing key environmental issues. Because these domains are not mutually exclusive and a CBM initiative may focus on more than one of these domains, a Venn diagram was used to illustrate these domains (Figure 4.6).

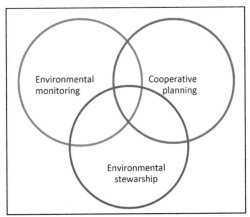

Figure 4.6 Typologies of CBM initiatives
Source Wehn et al. (2015a)

The mission of Grip op Water Altena describes this initiative as "a place (on- and offline) where collected observations, knowledge and warnings are shared, where bottlenecks and measures are constructively discussed along short communication lines and where it is clear which actions are taken by which party". This mainly falls within the domain of cooperative planning. However, the vision of Grip op Water Altena is to support creating a situation in which "the municipalities, water authority, citizens and farmers understand each other's interests and ways of working and are *together responsible for* limiting the damage by pluvial flooding in urban and rural areas". This vision calls for shared responsibilities for managing the issue of pluvial flooding, and thus falls within the environmental stewardship domain. Therefore, the domain of this initiative is both cooperative planning and environmental stewardship.

As discussed in the baseline analysis of this case, the Dutch water management system is a well-functioning system in which authorities are in charge of keeping the citizens safe against floods and citizens have a high level of trust in the authorities to do so. Given this highly institutionalized system of water management, with little or no provision for public participation, it is hard to imagine how the shared responsibility for managing pluvial flooding in Altena can be implemented. As stated by Murphree (2000), authority and responsibility are linked "when they are de-linked and assigned to different institutional actors both are eroded. Authority without responsibility becomes meaningless or obstructive; responsibility without authority lacks the necessary components for its efficient exercise" (Murphree, 2000, p. 4).

Geographic scope of Grip op Water Altena

As explained in the background of this case study (Section 4.1), Grip op Water Altena is located in 'Land van Heudsen en Altena'. This is a river island located in the estuary of the rivers Rhine and Meuse. It is enclosed by the rivers Boven Merwede (north), Afgedamde Maas (east) and Oude Maasje/Bergse Maas (south) and by the region of De Biesbosch (west). Inhabitants living in land van Heudsen en Altena form the potential pool of participants in this CBM (i.e. roughly 55000 people).

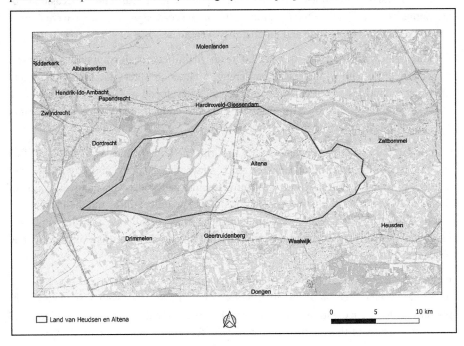

Figure 4.7 Geographic location of the Dutch Case Study
Source: Author

(Non) Participant groups in Grip op Water Altena

During the baseline analysis, a number of stakeholders that were relevant for Grip op Water Altena aims and activities were identified. The result of this analysis was presented in section 4.2.2.1 above. The aim of this section is to provide an overview of the stakeholder groups that in practice have participated in this CBM. This includes both stakeholder groups that participated in the collaborative process of designing the functionalities of Grip op Water Altena (i.e. co-design process), as well as end users of its tools and products.

The majority of the participants in the co-design workshops of Grip op Water Altena were community members, many of whom also represented environment-related NGOs. Representatives from the Regional Water Authority Rivierenland and the three municipalities of Werkendam, Woudrichem and Aalburg (and at later stages representative of the merged municipality of Altena) participated in almost all co-design workshops. KNMI, amateur weather networks, farmers association (ZLTO), housing association (Meander), and some local business owner were also among the participants in the co-design workshops; however, their participation was either occasional or discontinued after a while. Overall, participants in the co-design workshops were mainly citizens of a high average age and the majority were male.

All three interviewees from the GT2.0 team identified local community members (especially those with an interest in gardening, weather or water), the municipality and the water authorities as the end-users of the tools developed in Grip op Water Altena. However, the interviewees did not expect citizens living in apartments and the rural community (e.g. farmers) to actively use the Grip op Water Altena tools.

Comparing the actual participant groups in Grip op Water Altena with the identified stakeholders in the baseline of this case study shows that National and provincial level actors (e.g. Rijkswaterstaat, Ministry of Infrastructure and Water Management, Union of Regional Water Authorities, the Association of Dutch municipalities and the Province of Noord-Brabant) are among the stakeholders who were not involved in designing the functionalities of this CBM and are currently not among the end-users of its tools and products.

In addition, Google Analytics was used for analyzing the number of active users of Grip op Water Altena Web-platform. In this analysis, unique users who visited the web-platform at least once within a 28-day period were assumed as active users. Figure 4.8 shows the results of this analysis that was done for a one year time period from 1st of December 2018 until end of November 2019. As the figure shows, the number of active users of Grip op Water Altena web-platform fluctuated throughout this time period, but it mainly started increasing from June 2019 and reached a maximum of 41 active users in November 2019. Due to the fact that no registration is required for using the web-platform, it is not possible know which stakeholders are among these active users, but the chances are high that a few of them are GT2.0 team members.

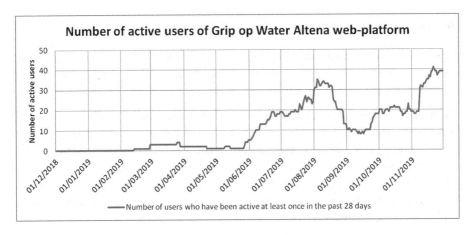

Figure 4.8 Number of active users of Grip op Water Altena web-platform

Efforts required to participate in Grip op Water Altena

Participation in Grip op Water Altena can be both online and offline. Online participation mostly requires access to internet, as well as a phone or computer. Participants need to take the time to post information (e.g. about measures taken in their garden), read the available information online, or promote the activities of the CBM via social media. Offline participation relates to participation in meetings, outreach event or campaigns, assistant with organizing such events, and the efforts needed for contacting and recruiting more members. This form of participation also requires time commitment, as well as travel expenses (if any), however, the future sustainability plan of this CBM has provisions for covering these costs (see section 4.3.3). Unlike several CBMs that identified knowledge requirements as a barrier for participation (Bartonova et al., 2016; COBWEB Consortium, 2015b; Novoa & Wernand, 2013), involvement in Grip op Water Altena did not require much specialized knowledge from citizen's side and therefore this was not elicited by any of the interviewees as a barrier for participation.

Nevertheless, interviewees from the authorities mentioned that they need to provide technical expertise regarding the topic of water management, and also convert the decisions made in the participatory meetings into resources and actions needed in their organizations. Moreover, they might need to communicate about these efforts and related challenges within their organization, or externally with other organizations. This indicates that additional efforts such as time and availability of staff members at the Regional Water Authority Rivierenland and the municipality of Altena are needed for participation of these stakeholders.

Support offered for participation in Grip op Water Altena

Support offered for participation by Grip op Water can be divided into three broad categories; organizational support, financial support, and the technical support required for developing the tools in this CBM.

Organizational support include scheduling, setting up, hosting and moderating the participatory meetings and outreach events, as well as setting up and maintaining social media accounts. The costs of organizational support were covered by the Ground Truth 2.0 project, and from 2020 the municipality of Altena and the Regional Water Authority Rivierenland will provide subsidies for covering these costs. Managing the social media accounts (e.g. the Facebook and twitter account) was done by the Ground Truth 2.0 project partners, but this was handed over to the CBM members.

Tools in this CBM includes development of the web-platform and integrating the data, maps and information available on the platform, as well as the development of two surveys for collecting information about gardens and past floods (2014 and 2015). The costs of designing and maintaining the Grip op Water Altena web-platform was provided by the Ground Truth 2.0 project and future maintenance costs of this web-platform will be covered by the municipality of Altena and the Regional Water Authority Rivierenland. A lot of the data and information available on the web-platform was provided by the stakeholders and especially the Regional Water Authority Rivierenland and the municipality of Altena. Moreover, the required technical support for linking databases was provided by the Ground Truth 2.0 partner organizations and the Regional Water Authority Rivierenland.

Pattern of communication in Grip op Water Altena

During the baseline interviews in this case study, the interviewees were asked to indicate their preferred channels for communicating about pluvial flooding. This question was designed to help prioritize the preferred communication channels from the users' point of view. Section 4.2.2 summarized the results of this baseline assessment. One of the findings of this assessment was that unless there is an actual problem or an emergency situation, there is little or no communication with municipalities and the water authorities. Moreover, community members mostly discuss about this topic in informal settings and face-to-face. Phone calls or SMS and emails are the most frequently mentioned communication channels and these were mostly used to communicate with the authorities in case there is a problem or complaint (see Figure 4.4).

In the second phase of the interviews, interviewees were asked to identify the communication channels, which they have actually used to participate in different activities of Grip op Water Altena, and the extent of use of each channel. Figure 4.9 summarizes the results of this inquiry. Several channels were used for communication in

Grip op Water Altena, however, emails and face-to-face communications were identified as the most frequent channels of communication in this CBM.

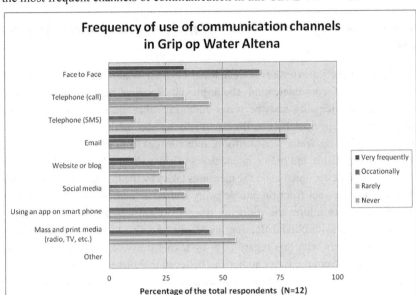

Figure 4.9 Frequency of use of different communication channels in Grip op Water Altena

Interviewees were also asked to indicate for what purpose they have used each communication channel. Emails were mostly used for coordination purposes (e.g. invitation to participatory meetings, setting up appointments), for exchanging information and also for sharing feedbacks and results of the CBM activities. Face-to-face communication channel corresponds to the participatory and bilateral meetings and also outreach events that meant to promote Grip op Water activities and recruit members. Telephone calls were mainly used for short bilateral discussions between members. Moreover, the CBM members used the Grip op Water website for sharing and viewing information e.g. about the measures taken in the area by different stakeholders, or information about water levels in the area. Other less frequently used channels included the Facebook and Twitter accounts that were mainly used for promoting the CBM activities and results and a WhatsApp group with eight members (besides the GT2.0 team members) that was mainly used for coordination among the members. Lastly, print material, flyers and banners were also used in the participatory meetings and outreach events.

Overall, Grip op Water Altena has created a number of bi-directional communication possibilities among the stakeholders. However, an online mechanism for interactive communication is lacking and this pattern of communication is mainly limited to the offline mode and via face-to-face interactions among the CBM members.

Change in methods of communication and participation because of Grip op Water Altena

The majority of the interviewees believed that Grip op Water Altena has provided an alternative way of communication about pluvial flooding between the community members, the municipality and the water authorities that did not exist before. These communications happened both offline (e.g. during the participatory meetings), and online via the Grip op Water web-platform. Interviewees from the community members believed that the offline and online interaction in Grip op Water has provided them with an opportunity to better understand the measures taken by the municipality and the water authorities. Interviewees from the municipality and water authorities saw this initiative as an opportunity to inform the residents about the measures that they have taken for reducing problems with pluvial flooding in Altena, and also to ask for collaboration of community members with this regard (e.g. by taking measures in their own gardens). Both interviewees from the water authorities believed that there were much more offline interactions in the beginning and this decreased towards the end of the Ground Truth 2.0 project. Lastly, an interviewee from the municipality considered the two surveys that were conducted as a part of this initiative as a complementary form of communication and information sharing between citizens and the authorities.

Nevertheless, none of the interviewees from the community members could identify an example of a mechanism by which Grip op Water has enabled them to take part in decision making processes related to the management of pluvial floods. Moreover, interviewees from the municipality and the water authorities indicated that they have already been involved in making decisions about this issue and believed that Grip op Water did not change anything in that regard.

Overall, comparing these results with the baseline situation of this case, and the 'communication and decisions modes' identified by Fung (2006) and Wehn et al. (2015b) shows that Grip op Water Altena has facilitated communication between different stakeholders. This is done by providing the stakeholders with possibilities for learning about each other's activities and wishes, sharing information, expressing preferences (e.g. in face-to-face meetings) and developing preferences through discussions and information exchange. Nevertheless, this CBM has not created a change in 'modes of decision making' or possibilities for collaboration of citizens in decision making processes about management of pluvial floods (Fung, 2006).

4.3.3 Power dynamics in Grip op Water Altena

This section is dedicated to presenting the results of an analysis of the external and internal power dynamics in Grip op Water Altena. This is directly linked to the third research question of this study, i.e. who controls and influences Grip op Water Altena and how?

Change in the social, institutional and political context of the case

Section 4.2.1 provided an overview of the social, institutional and political context of the Dutch case study. Since this information was generated at the start of the establishment process of Grip op Water, the researcher was interested in capturing any changes in those contextual settings that affected the establishment process or results of the CBM.

The most tangible institutional change in this case was the merger of the municipalities Werkendam, Aalburg, and Woudrich and formation of the new municipality of Altena as of 1st January 2019. The GT2.0 team members believed that this merger of the municipalities did not affect Grip op Water Altena, because people involved in the CBM were more or less the same and their point of view about the initiative did not change. However, it was mentioned that during the lead up to the merger, their representatives had little time for Grip op Water.

There was also a recently introduced policy that aligns very well with the aims of Grip op Water Altena and targets active participation of citizens in climate proofing urban areas. This policy is in the form of a subsidy called 'Subsidie Klimaatactief' and is provided by the Regional Water Authority Rivierenland. This subsidy is given to residents that take measures for making their garden, house or street climate-proof. The subsidy amount is 35% of the eligible costs, which can be between 1,000 to 15,000 Euro. Replacing garden tiles with grass or plants, making a green roof, or installing a rain barrel are examples of eligible measures for receiving this subsidy[15].

Some community members identified a change in the implementation side of water management practices in Altena. They believed that during the past few years, authorities have taken quite a few measures against pluvial flooding. These were mainly structural measures that aimed at improving the drainage system in the region of Altena. Changing the road surface, digging new canals and renewing the existing drains were examples of these changes.

---

[15]For more information about 'Subsidie Klimaatactief' see: https://www.hohohoosbui.nl/subsidie-klimaatactief/

Establishment mechanism of Grip op Water Altena

Firstly, it is important to highlight that the need of establishing Grip op Water Altena was initially proposed by the Ground Truth 2.0 project and the required resources for establishing this CBM were made available through this project. Therefore, the initiation of establishing Grip op Water Altena is project-driven or supply-driven, and not demand-driven. This is particularly important to note, because similar to any other project-driven CBM, Ground Truth 2.0 had pre-defined timeframe and budget for establishing a CBM in its Dutch case study. It received funding from the European Union and had certain obligations towards its funder (e.g. to establish six CBMs using a co-design approach in pre-defined countries). Moreover, partner organizations and team members in this case had certain research interests and expertise, factors that inevitably shaped the establishment process, functioning and results of Grip op Water Altena. For example, the core focus of HydroLogic Research (i.e. the technical partner of this case) is innovative technical solutions for water management problems and team members from this organization were hydrologists and ICT experts. Similarly, the core focus of IHE Delft is research on the topic of water, but the team members in this case included both social scientists and people with technical background (e.g. GIS), who had an interest in citizen science. Regardless of the fact that the inputs from the co-design process shaped the functionalities of Grip op Water Altena, the design of the co-design methodology and instructions for holding the co-design meetings was mainly informed by the existing knowledge and expertise at IHE Delft. Moreover, a number of existing technological solutions and products from HydroLogic Research were integrated in the technical design (see Table 4.2).

In theory, the process used for establishing Grip op Water Altena indicates a 'co-created' or 'co-designed' model of establishing a CBM initiative (Conrad & Hilchey, 2011; Haklay, 2015; Shirk et al., 2012; Wehn et al., 2015a). Nevertheless, the researcher was interested in understanding the perception of the interviewees about the establishment process of this initiative.

The majority of interviewees from the CBM members mentioned that Grip op Water Altena was established using a co-design method, and indicated that they had the chance to influence the design and functionalities of this initiative. Nevertheless, two interviewees from community members believed that the establishment process of Grip op Water Altena followed a top-down model, in which authorities in charge of water management had a stronger voice. These interviewees believed that although citizens were consulted in the process, the extent of their involvement was limited.

Two of the three interviewees from the GT2.0 team, believed that Grip op Water Altena was co-designed by a group of interested stakeholders who had the chance to influence its design and functionalities. However, one interviewee described the establishment process of this CBM as a top-down model that was initiated and mainly driven by interests and wishes of the researchers. This interviewee indicated that the process was research-driven and "we did not co-design the co-design process"[16].

Change in access to and control over data because of Grip op Water Altena

The website of Grip op Water Altena integrates weather and water information from various sources. Available information on this website includes physical characteristics of the case study area, precipitation information (including both rainfall data and forecasts), pluvial flooding observations by citizens, information about past flood events of 2014 and 2015 and risk of flooding in different areas, as well as measures taken by different stakeholders to limit the damage from pluvial flooding in Altena (Table 4.3). Information on the Grip op Water Altena website is mainly provided by the Regional Water Authority Rivierenland, HydroNET[17], and a few other sources. However, citizens can also contribute to the information to this website. More specifically, citizens can contribute to observations of pluvial flooding (submitted using a form that can also include a picture), measures taken to limit the damage from pluvial flooding (i.e. also submitted through a form), as well as memories from the flood event of 2014 and 2015. At the time of conducting this research, the citizen-contributed data on the website of Grip op Water Altena was very limited and included three measures taken in private residential buildings for reducing the risk from pluvial flooding, seven pluvial flooding observations, and five memories and pictures of flooding in 2014 and 2015.

In addition to the above mentioned information, two surveys were designed and conducted as a part of activities in this CBM. The first one was a garden survey that was conducted in October 2018, with the aim of gaining more insights into the reasons why residents of Altena have a green or paved garden. This survey was completed by 232 respondents and provided valuable information about the status of gardens across the case study area. Figure 4.10 provides the geographic distribution of the respondents in this survey[18]. The second survey aimed at collecting memories and information about the floods in July 2014 and August 2015. Thirty respondents participated in this survey. The analysis of the results showed that people from Woudrichem expect more nuisances from pluvial flooding than people from other towns. Moreover, measures that were taken by the municipality and the Regional Water Authority Rivierenland are not always noticed by citizens.

---

[16] NL-02-01

[17] A decision support system developed by Hydrologic; for more information see https://www.hydronet.nl/over-hydronet/

18 For more information about the results of this survey visit: http://altena.gripopwater.nl/oudere-inwoners-van-altena-hebben-de-groenste-tuinen/ (in Dutch)

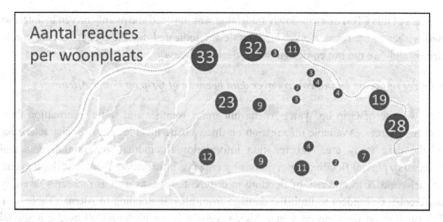

Figure 4.10 Spatial distribution of respondents to the garden survey in Grip op Water Altena[19]

[19] Source: http://altena.gripopwater.nl/oudere-inwoners-van-altena-hebben-de-groenste-tuinen/

Table 4.2 Overview of weather and water information available on Grip op Water Altena website

Category	Data/information	Description	Source of data/information
Physical characteristics of the area	Surveillance areas	Information about monitoring areas in Altena, which are regulated by weirs and pumping stations	The Regional Water Authority Rivierenland
	Status of surface water	Water level of the water bodies in Altena	The Regional Water Authority Rivierenland
	Elevation	Topographic and water level information of Altena	Esri Nederland and AHN
Precipitation	Current rainfall	Accumulated precipitation data for the past three days in Altena	HydroNET
	Expected rainfall	Predictions of rainfall for the next ten days in the Central Netherlands region	KNMI and HydroNET
	Precipitation deficit	Trend of precipitation deficit (mm) over time, averaging over 13 stations throughout the Netherlands	KNMI
Observations	Pluvial flooding observations	Reports submitted by citizen using a form. This Form is used to report observations about problems with pluvial flooding in Altena	Citizens
Flooding	Flood in 2014	Map and rainfall data of the July 2014 flood in Altena	HydroNET
	Flood in 2015	Map and rainfall data of the August 2015 flood in Altena	HydroNET
	Memories of flooding in 2014 and 2015	Memories and pictures of flooding in 2014 and 2015 shared by citizens	Citizens
	Climate atlas of floods in Rivierenland	A map that indicated vulnerable locations to flooding in the management area of the Regional Water Authority Rivierenland, based on modeling a heavy summer rain	The Regional Water Authority Rivierenland, Deltares and HydroLogic Research
	Maximum water depth with intense precipitation	A map that shows maximum water depth that can occur due to short-term intense rainfall of 70 mm in 2 hours. This data is available for the whole country	Developed by Deltares and published in the Climate Effect Atlas
Measures	Measures taken to limit the damage from pluvial flooding	Measures taken by residents of Altena, the Altena municipality and the Regional Water Authority Rivierenland to limit the damage from pluvial flooding in Altena	Citizens, the municipality of Altena and the Regional Water Authority Rivierenland

The results of interviews demonstrated that change in access to data because of Grip op Water Altena is perceived differently by the GT2.0 team and the local stakeholders.

The GT2.0 team members believed that Grip op Water website integrates a lot of information from different sources and also provides valuable information about what measures have been taken by which stakeholder. It also provides information about the history of flooding in the area and an easy way of finding the areas of high pluvial flooding risk.

Community members, the municipality and the water authorities unanimously believed that Grip op Water Altena has created little or no change in their access to and control over data and information. Interviewees from the water authorities and the municipality believed that the available data on the website of Grip op Water Altena is mainly provided by the Regional Water Authority Rivierenland and therefore it has not changed their access to, and control over, data and information. In this regard, one of the interviewees from the Regional Water Authority Rivierenland mentioned that if new data would be generated by citizens in this initiative, we would use that data, however this has barely happened[20]. The majority of the interviewees from community members indicated that their access to data and information has not changed because of their participation in this initiative and the available information could have been accessed from other sources. Interviewees from community members also mentioned that they have not used the data and information provided on Grip op Water Altena website. The only exception was one interviewee who mentioned she has used the CBM website to see what measures have been taken to prevent flooding[21].

Change in the authority and power of different actors because of Grip op Water Altena

As a part of this study, the researcher was interested in understanding any change in the levels of authority and power of different stakeholders as result of their participation in Grip op Water Altena. In order to do so, both the GT2.0 team members and the CBM members were asked to what extent, if any, they think their influence in decision making processes regarding management of pluvial flooding in Altena has changed because of their participation in this CBM.

Interviewees from the GT2.0 team members believed that Grip op Water Altena has provided them with an opportunity to raise awareness within the municipality and among the community members on how they can reduce the risk of pluvial flooding using a collaborate approach and with the help of all involved stakeholders. However, it was also mentioned that there has been a broader movement about the topic of pluvial flooding in

---

[20] NL-02-07
[21] NL-02-09

the Netherlands and more people are aware of this issue, but that is not only because of Grip op Water Altena.

Interviewees from the municipality and water authorities indicate that they were already a part of decision making processes, related to their role and function at their organization and Grip op Water Altena did not change this. The majority of the CBM member did not perceive a major change in their influence on decision making processes. They however mentioned that participation in Grip op Water Altena has provided them with an opportunity to learn about the topic of pluvial flooding and has raised their awareness about measures taken by different stakeholders (especially the municipality and water authorities) for reducing the risk of pluvial flooding. The only exception was one interviewee who believed because of the enthusiasm that he has about Grip op Water Altena, he may be able to communicate about and influence the opinion of other community members.

Comparing these results with the baseline situation of authority and power in this case, and the identified levels of influence on decision making processes by Fung (2006) suggests that Grip op Water Altena has not changed the level of authority and power of different stakeholders. Rather it has created an alternative possibility for all stakeholders to exert communicative influence via dialogues and information exchange about the issue of pluvial flooding in Altena.

Revenue streams of Grip op Water Altena

The results of the interviewees with CBM members and the GT2.0 team members demonstrate a common understanding among the majority of interviewees about the future revenue streams for this initiative.

Based on the future sustainability plan of Grip op Water Altena, this CBM will be a working group under an NGO called Agrarische Natuur Vereniging (ANV) Altena; (Kersbergen et al., 2019). The organizational support required for running future activities and meetings in this initiative will be provided by ANV and the Regional Water Authority Rivierenland.

The future costs of running this CBM will be covered by a 'government sponsorship' model (Gharesifard et al., 2017, 2019b). This includes the costs of hosting the website, subscription to products such as HydroNET, promotion costs, and venue for future meetings. The HydroNET and ArcGIS-online subscriptions are current running subscriptions at the Regional Water Authority Rivierenland. The municipality of Altena and the Regional Water Authority Rivierenland will cover the costs of hosting the Grip op Water Altena website, and they will help with providing a venue for future meetings in this CBM.

4.3.4 Technological choices for Grip op Water Altena

The content of this section, aims to answer the fourth research question of this study for Grip op Water Altena (i.e. How effective and appropriate are the choices and delivery of the selected technologies of Grip op Water Altena?). This is done by providing an overview of the technological components used in Grip op Water Altena, assessment of accessibility of these technologies and how these relate to existing infrastructure, as well as discussions about included and excluded groups as result of technological choices.

Technologies used in Grip op Water Altena

Technical design of all CBMs in Ground Truth 2.0 was informed by the required functionalities in each case. Required functionalities of each CBM were identified using a functional design process. This process aimed at translating users' requirements to functionalities that can be then used by CBM members to interact, communicate and exchange information (Alfonso et al., 2017a). In order to do so, a Story Mapping approach was used to capture the needs and wishes of CBM members. These needs and wishes were captured in the co-design meetings using a 'story card' that captures the type of user, desired functionality and the perceived added value from that functionality (Alfonso et al., 2017b). Figure 4.11 presents the generic format of the cards used for capturing the user stories.

As a *<type of user>*

I want *<to do something>*

so that *<some value is added>*

Figure 4.11 Generic format of the user story cards
Source: (Alfonso et al., 2017b)

A Generic Story Map was developed that covers a wide range of functionalities for all CBM initiatives (for further details see: Alfonso et al., 2017a). This Generic Story Map was then used as a reference for identifying specific functionalities of the CBMs in Ground Truth 2.0 project.

The story map analysis in Grip op Water Altena resulted in identifying two main categories of functionalities, which then guided the technical design of this CBM; (1) a platform to access and share water and weather-related information, and (2) channels for communication among different stakeholders (Giesen, 2018). Figure 4.12 shows the story map for Grip op Water Altena.

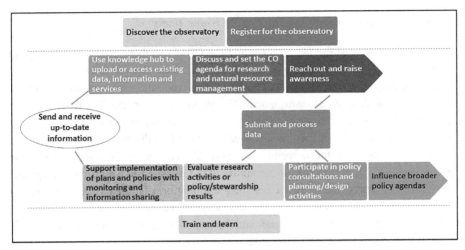

Figure 4.12 Story map of Grip op water Altena
Source: Giesen (2018)

The main technological component of Grip op Water Altena is a web-platform[22]. This web-platform contains a number of interactive maps with water weather and information (mainly from external data sources). These maps are integrated from various open and private sources, including HydroNET, the Regional Water Authority Rivierenland and other sources (Table 4.3). These maps are accompanied by supplementary information and using the ESRI Story Maps interface (Giesen, 2018). Moreover, static information about the measures that can be taken for reducing the risk of pluvial flooding is compiled from different sources and shared through the web-platform. In addition, the web-platform of Grip op Water Altena enables submission of relevant information and observations via dedicated Google Forms and Esri Survey123. Reports about nuisance from pluvial flooding and information about measures taken in private gardens against pluvial flooding are examples of these observations and information. A Google Maps plug-in is used to visualize these reports on maps. Annex 6 presents a screen shot of Grip op Water Altena main page, as well as examples of forms and maps on this platform.

The second category of functionalities (i.e. communication and interaction among stakeholders) is supported both offline (e.g. in face-to-face meetings) and also via a number of external communication tools, such as a WhatsApp group, a Facebook page, and a Twitter account are used for communication among the CBM members and outreach to a wider public.

---

[22] http://altena.gripopwater.nl/

Accessibility of technologies used by Grip op Water Altena

The interviewees from the GT2.0 team and the CBM members believed that Grip op Water web-platform can facilitate a two-way communication between the local authorities the larger community of residence in Altena, and especially those with an interest in the topic of water management. All interviewees from the GT2.0 team and the CBM members were of the opinion that Grip op Water Altena web-platform is easily accessible and mentioned that it is designed in a user friendly and logical way. In this regard, one of the GT2.0 team members mentioned that because the composition of the people involved in defining the functionalities of the website had a high average age and were less tech-savvy, therefore the resulting web-platform is fairly easy to use for an average user. It was however mentioned that some of the maps on the website are a bit more complicated for an average user, and the use of these maps may require a certain level of knowledge or technical expertise.

Included and excluded groups resulting from technological choices

Overall, there is a very good match between the technological choices in Grip op Water Altena and the existing infrastructure, as well as social and technological capabilities in this case study.

Developing a user-friendly web-platform in the local language, with an easy to use interface, form-based observation submission options, and widespread use of visual material (e.g. maps and pictures) enables a large proportion of residence in Altena to use this web-platform.

Given the results of the baseline analysis in this case that indicates a very large proportion of the Dutch citizens are active technology users, and also the possibility of offline interactions during the face-to-face meetings for those who lack basic digital skills, this initiative does not exclude participation of a major part of the population in Altena.

4.3.5 Results of Grip op Water Altena

As a part of this research, interviewees were asked to identify expected and realized outputs (direct products) of Grip op Water Altena. They were also asked to express their opinion about the realized outcomes (actual short-term or incidental changes) that have happened because of Grip op Water Altena, and the outcomes that can be expected to happen in the near future. This is directly linked to the fifth research question of this study (i.e. what are the expected and realized outputs, and interim outcomes of the CBM initiative?). Impacts take a long time to materialize and their study is out of the scope of this research. Nevertheless, interviewees were also asked to express their expectation of long-term changes that may happen in the future as a result of Grip op Water Altena. The findings reported in this section summarize the expected and realized outputs, outcomes and impacts of Grip op Water Altena. It should be mentioned that although interviewees were explicitly asked to think of both positive and negative outputs, outcomes and impacts, uncertainty about future outcomes and impacts and a certain level of 'social desirability bias' (Fisher, 1993), because of their affiliation with the CBM, may have influenced the responses.

Outputs of Grip op Water Altena

The GT2.0 team members, authorities and community members had slightly different views about the expected outputs of Grip op Water Altena, but they were largely in agreement about its realized outputs.

Establishment of a CBM initiative that engages different organizations and local community members in collection and sharing of weather and water data and information was one of the main expected outputs for the GT2.0 team members. Data about measures taken by different stakeholders for reducing the risk of pluvial flooding and information about water storage capacity of private gardens in the case study area were mentioned as examples of these water and weather related data and information. This information was expected to be shared via a web-platform that is used by authorities and a large number of community members. Moreover, the GT2.0 team members expected to facilitate communication about the issue of pluvial flooding by establishing an institutionalized communication procedure. Interviewees from this group believed Grip op Water Altena was successful in developing a web-platform for sharing data and information and engaging the municipality of Altena, the Regional Water Authority Rivierenland and a small number of community members. However, the data and information sharing on the platform was described as mainly one directional and from authorities to citizens. It was mentioned that a very limited amount of data and information was shared by citizens, and this was mainly limited to the results of the two aforementioned surveys.

The Regional Water Authority Rivierenland was expecting to create more awareness and also obtain area-specific information that can help improve decision making processes regarding management of pluvial floods, in a more collaborative way. The interviewees from this organization believed that Grip op Water Altena has been successful in creating a web-platform and facilitating occasional interactions between citizens and several organizations. Nevertheless, this CBM has not so far been successful in producing area-specific information that are used by authorities for improving management of pluvial floods in Altena.

The interviewee from the municipality mentioned that he did not expect a specific output from Grip op Water Altena; however, he described the website of this CBM as a good communication tool that contributed to awareness raising at a small scale, and facilitated creating contact with other organizations.

Interviewees from the local community identified quite a diverse range of outputs that they expected to see as a result of Grip op Water Altena. Minimizing water nuisance in the case study area by taking measures against pluvial flooding, communicating with authorities about existing issues, having a say in decision making processes regarding management of pluvial floods, sharing information and creating a support base for the authorities were among these expected outputs. These interviewees also expressed their opinion about the actual outputs of Grip op Water Altena. It was mentioned that through communication and direct contact between different stakeholders in Grip op Water Altena, authorities are now more aware of the problems. Moreover, some information is shared via the CBM website; however interviewees believed that this information has resulted in limited awareness raising and no practical measures have been taken with this information.

Outcomes of Grip op Water Altena

Most interviewees from all groups believed that the realized outcomes of Grip op Water Altena are (yet) limited; however they identified a number of outcomes for this CBM that were mainly societal and governance-related. Creating a small community of stakeholders around the topic of pluvial flooding, awareness raising about participatory approaches for reducing the risk of pluvial flooding (mainly within this community), and creating a new way of communication and interaction between municipality, water authorities and citizens were the main realized outcomes of Grip op Water Altena. Similarly, the majority of interviewees did not expect major short-term changes resulting from this CBM in the near future; however, some interviewees mentioned that this CBM has the potential to contribute to more awareness raising, data sharing, and better communication and interaction among the stakeholders. In this regard, the interviewee from the municipality of Altena mentioned; "maybe we can have a risk dialog via Grip op Water Atena"[23].

Expected impacts of Grip op Water Altena

Although a number of interviewees from all groups were skeptical about the future impacts of Grip op Water Altena, or described it as largely unknown, others identified possible future impacts for this CBM. The identified impacts were mainly environmental, societal and governance-related. It was however mentioned that realization of these expected impacts depends on the future uptake of the activities in Grip op Water Altena.

An interviewee from the Regional Water Authority Rivierenland mentioned that Grip op Water Altena, along with a series of other efforts at the local and national level, has

---

[23] NL-02-10

contributed to planting a seed for participatory approaches for water management in Altena. Some interviewees believed that this participatory approach will result in closer collaboration among different stakeholders and will eventually help with 'getting water more under control' and creating 'more space for water'.

Moreover, it was also expected that improved communication among stakeholders will result in increased awareness, trust and transparency among stakeholders. A few interviewees expected this to contribute to a change in attitude towards environmental stewardship. More specifically, these interviewees expected that Grip op Water Altena contributes to an increase in active participation of community members in reducing the risk of pluvial floods by taking measures themselves, instead of only criticizing the authorities in charge.

4.4 DISCUSSION

The findings presented in sections 4.2 and 4.3 summarized the results of a systematic analysis of the baseline situation of the Dutch case study, as well as the establishment process and results of the CBM in this case i.e. Grip op Water Altena. The findings presented in this chapter are informed by the results of the two phases of empirical research in which the CPI Framework was used as a guiding frame for data collection, analysis and reporting. The findings from context analysis, process evaluation and impact assessment of Grip op Water Altena help answer research questions 1 to 5 of this study by unpacking the complex processes involved in the establishment and functioning of this CBM. The discussions presented in this section are framed around these five questions and aim at answering these questions in a succinct and accessible way, while linking the findings to relevant literature and current debates about establishment and functioning of CBMs. Answers to these five questions help depict the meaning of community in the context for Grip op Water Altena by clarifying its objectives and actor-specific goals, technological components, participation processes and power relationships in this CBM.

RQ1. What are the overarching objectives and actor-specific goals of Grip op Water Altena and to what extent does the design of this initiative help achieve those goals/objectives?

Grip op Water Altena's vision and mission portray a CBM that aims at facilitating cooperative planning and environmental stewardship around the topic of pluvial flooding in Altena.

Using a co-design approach, the members of this CBM jointly defined five rather ambitious and broad objectives for this initiative (Table 4.1). Throughout the establishment process, a shift of attention happened in achieving these objectives. More specifically, the first and the second objectives that aimed at facilitating exchange of observations and supporting short communication lines got out of focus. At the same time, Grip op Water Altena focused more on setting up a knowledge platform for exchanging perspectives and tips to take measures against damage from pluvial flooding, and also supporting open and constructive dialogue between all involved stakeholders; especially citizens, the Regional Water Authority Rivierenland and municipality of Altena.

Even though the five co-designed objectives of Grip op Water Altena align with the identified actor-specific goals in this CBM, the aforementioned change in focus of the objectives resulted in less focus on collecting and sharing data about weather and water systems, and also supporting short communication lines. While collecting and sharing citizen-contributed observations was mainly a desire from the water authority and municipality, establishing short communication lines was a specific wish from the local community members who participated in the co-design process.

This change in focus was influenced by a number of factors, including the small number of CBM members, the existing awareness gap regarding water management issues among Dutch citizens, high level of trust in authorities among Dutch citizens, and the highly institutionalized top-down water management system in the Netherlands, that does not envision or allow for public participation. As the baseline analysis of this study and previous research (e.g. Kaufmann et al., 2016; OECD, 2014) indicate, there is an existing awareness gap within the Dutch community regarding water management practices. Therefore, the majority of Dutch citizens does not perceive pluvial flooding a major issue and trusts the authorities to manage this issue. This might have been a contributing factor to the low level of engagement of community members with Grip op Water Altena, which in turn made it difficult for this CBM to generate a large number of citizen-contributed observations. Moreover, well-established water management practices at the Regional Water Authority Rivierenland and municipality of Altena did not allow for creating alternative short communication channels and agreeing on possibilities for data collection. This is in line with findings of Hecker et al. (2019) that identified difficulty of alignment with existing official structures and existing ways of working as a challenge for citizen science projects.

90

Linked to this point, during the functional design of Grip op Water Altena a paradox was identified regarding the focus of Grip op Water Altena on environmental stewardship and the fact that this CBM is interacting with a highly institutionalized system of water governance in the Netherlands (Alfonso et al., 2017a). Establishing a CBM with the aim of environmental stewardship and shared responsibilities in a highly institutionalized system of environmental governance is very challenging, if not impossible. Environmental stewardship calls for shared responsibilities and as Murphree (2000) states, exercise of responsibility requires authority. In the case of Grip op Water Altena, the Regional Water Authority Rivierenland and the municipality of Altena have the official mandate for managing pluvial flooding in Altena and citizens would be expected to share this responsibility without any authority. In order for such a CBM to successfully achieve its ambitions, there is a need for an institutional change that accommodates for a stronger role for public participation in water management practices.

RQ2. Who participates in Grip op Water Altena and how?

Participation in Grip op Water Altena can be divided into two phases, in which the composition of involved stakeholders remained mainly the same. The first phase was the co-design phase in which this CBM was established and its functionalities were defined. The second phase is the current phase in which members of Grip op Water Altena can utilize the co-designed functionalities for participation in this initiative. Overall, this CBM allows for both offline and online participation. Offline participation is mainly via face-to-face meetings and outreach events, and online participation mainly happens through the Grip op Water Altena web-platform. Water authority Rivierenland and the Municipality of Altena are the two government organizations that have been involved since the beginning of the co-design process of Grip op Water Altena. Moreover, a small number of interested community members (often of higher average age ranges, and mainly male) also participated in the two aforementioned phases, some of whom also represented environmental NGOs.

Overall, Grip op Water Altena has created a number of two-way communication and information exchange possibilities among the Regional Water Authority Rivierenland, the Municipality of Altena and citizens. This created a change in communication modes by facilitating bi-directional communication between different stakeholders. However, an online mechanism for interactive dialogue between stakeholders is lacking and this pattern of communication is mainly limited to the offline mode and via face-to-face interactions among the CBM members.

Engaging a large number of participants in Grip op Water Altena proved to be difficult. The findings presented in this chapter show that the number of participants in the co-design process of Grip op Water Altena dropped over time. In principle, beyond the co-design group, anyone who knows about Grip op Water Altena can participate in the second phase and utilize the functionalities and information produced in this CBM. The number of active users of the web-platform of Grip op Water Altena has fluctuated over time, but did not exceed a maximum of 41 active users. Unlike several CBMs that identified knowledge requirements as a barrier for participation (Bartonova et al., 2016; COBWEB Consortium, 2015b; Novoa & Wernand, 2013), participation in the co-design process or using the web-platform of Grip op Water Altena did not require much specialized knowledge from citizen's side. In addition, organizational and technical support required for online and offline participation of interested members was provided and preferred communication channels were used for contacting local community members. There are however two possible explanations for the aforementioned low level of participation in Grip op Water Altena. The first explanation relates to the lengthy process of consensus building and establishment of Grip op Water Altena. The slow process of establishment has already been identified as a drawback of co-created CBMs (Bonney et al., 2009a; Frensley et al., 2017). Grip op Water Altena was not an exception and the lengthy process of consensus building, during which no tangible results or tools were produced, led to discontinued participation of a number of community members and organizations. The second explanation relates to the social and institutional settings of this CBM. The existing awareness gap within the Dutch community regarding water management practices (OECD, 2014), perceived low importance of the issue of pluvial flooding by citizens, and high level of trust in authorities to manage this issue may also have contributed to the low level of participation in Grip op Water Altena. Moreover, no pluvial floods happened since the inception of Grip op Water Altena and therefore the sense of urgency of the topic remained low among the local community members.

RQ3. Who controls and influences Grip op Water Altena and how?

The majority of interviewees agreed that Grip op Water Altena has been co-designed, and they have had the chance to influence the design of vision, mission, objectives and functionalities. Co-designed or co-created CBMs usually provides the parties involved with a more equal chance to influence the establishment processes (Conrad & Hilchey, 2011; Wehn et al., 2015a), as compared to top-down or bottom-up initiatives (Ciravegna et al., 2013; Conrad & Hilchey, 2011). Nevertheless, it is important to highlight the fact that Grip op Water Altena is a project-driven CBM. Onencan et al. (2018) argue that research projects are supply-driven and often for the purpose of satisfying particular organizational or project needs, or for matching budgetary requirements. The initial need for the establishment of Grip op Water Altena was based on the fact that Ground Truth 2.0 received funding from the European Commission to establish a number of CBMs in Europe and Africa. Regardless of the fact that these CBMs were co-designed and their objectives were co-created in consultation with local stakeholder, the initial idea for their establishment did not come from the local stakeholders and in that sense, they are not demand-driven CBMs. The GT2.0 team was in charge of designing the methodology for co-design and had more control over the process of establishment of this CBM. In addition, certain project requirements such as a pre-defined timeline, available resources and the initial framings of its focus on the issue of pluvial flooding (Ground Truth 2.0 consortium, 2015) influenced the objectives, establishment process, functioning and subsequently results of this CBM.

In comparison with the community members, authorities involved in Grip op Water Altena could exert more influence on certain processes and aspects of this initiative. For example, the Regional Water Authority Rivierenland and the Municipality of Altena were not in favor of creating short communication channels via Grip op Water Altena and preferred to use the existing channels for daily dialogue with citizens about pluvial flooding. In another example, the Municipality of Altena vetoed a number of possibilities for data collection in this CBM. The reason for this veto was the fact that the municipalities are the custodians of water management practices in urban areas, and the Municipality of Altena did not find the proposed possibilities for data collection particularly useful for improving their services. This aligns with findings of Cleaver (1999) and Newman et al. (2012) which suggest that governance of water and environment-related issues are inherently political and the existing power dynamics, competing interests and conflicting norms related to management of those issues should be accounted for.

In terms of change in access to and control over data, although Grip op Water Altena has slightly increased the access to weather and water data in Altena by integrating information from various sources, this is not considered a significant change in access to data for local stakeholders. This is due to the fact that Grip op Water Altena has generated little new data and information and most of the available information on the web-platform of Grip op Water Altena was already accessible through other sources.

Although Grip op Water Altena did not contribute to a change in the level of authority and power of different stakeholders, it provided the participants with alternative possibilities for exerting communicative influence (Fung, 2006) via dialogues and information exchange about the issue of pluvial flooding in Altena. The baseline analysis of this case shows that citizens communicate their concerns about management of pluvial flooding to authorities via different means, e.g. by writing a letter to municipality or water authority, by signing a petition or via social media. However, Grip op Water Altena provided its members with alternative possibilities of discussing these issues, e.g. via face-to-face interactions in the CBM meetings, or online and by submitting observations via the web-platform of Grip op Water Altena.

During the lifetime of Ground Truth 2.0 project, the financial means as well as the organizational support for establishing Grip op Water Altena, was provided by the project. The envisioned revenue streams for future sustainability of this CBM will be provided through government sponsorship (Gharesifard et al., 2017; Osterwalder & Pigneur, 2010), i.e. by the Regional Water Authority Rivierenland and Municipality of Altena, and the required organizational support will be provided by ANV. Moreover, the data and information shared via Grip op Water Altena platform is mainly provided by the Regional Water Authority Rivierenland. Regardless of the existing interest within the municipality and water authority for the continued activities of Grip op Water Altena, their official mandate, technical expertise and future financial sponsorship role shifts the balance of power in this CBM more towards the authorities.

RQ4. How effective and appropriate are the choices and delivery of the selected technologies of Grip op Water Altena?

The main technological component of Grip op Water Altena is a web-platform that supports access to, and possibilities for sharing water and weather-related information. Previous research, e.g. Gouveia and Fonseca (2008) and Newman et al. (2012) warned about inadvertently widening the 'digital divide' gap between those who own or adopt the technologies developments in a CBM and those who avoid it or lack the required skills to access those technologies. This was not the case for Grip op Water Altena. The CBM members evaluated the web-platform of this initiative as very accessible and user-friendly. The incorporation of visual materials such as maps and pictures makes the use of the website easier and observations by community members can be submitted in a form-based format that does not require much expertise. The platform is designed in the local language and therefore only excludes non-Dutch speakers who are not a large part of the local population.

Nevertheless, interactive communication possibilities are mainly external or offline and the platform does not support interactive online dialogues about the issue of pluvial flooding. Online interactive communication possibilities could have contributed to a better achievement of the second objectives of this CBM, i.e. supporting short communication lines, as well as its fourth objective i.e. supporting open and constructive dialogue between all involved parties. As argued by several STS scholars, e.g. Winner (1980), Mansell and Wehn (1998), MacKenzie and Wajcman (1999) and Bijker et al. (2012), technological developments are not only determined by technical possibilities for their creation, but also by societal needs, financial resources, political agendas and vested interests of developers and end-users. In the case of Grip op Water Altena, not including interactive communication possibilities in the technical development of the web-platform was mainly determined by inflexibility of the government organizations for accepting alternative channels of communication.

RQ5. What are the expected and realized outputs and interim outcomes of Grip op Water Altena?

Analysis of the outputs of Grip op Water Altena shows that although different stakeholder groups had slightly different views about the expected outputs of this CBM, they were largely in agreement about its realized outputs. CBMs can focus on both producing and measuring scientific results, as well as facilitating collaboration in environmental management (Conrad & Hilchey, 2011; Keough & Blahna, 2006). Grip op Water Altena mainly facilitated the collaboration and communication between local community members and authorities in Altena. Outputs of this CBM were mainly individual, societal and governance-related in nature. In practice, Grip op Water Altena provided online and offline possibilities for interaction and information exchange between stakeholders via a web-platform and face-to-face meetings. Available possibilities for interaction can be leveraged by different stakeholders to discuss issues and concerns, and also to exert communicative influence (Fung, 2006) on decisions regarding management of pluvial flood in Altean. Moreover, information sharing in face-to-face meetings and via the web-platform has led to an increase in understanding of CBM members about the measures taken by different stakeholders to reduce the risk of pluvial flooding and how this issue can be managed in a participatory way. However, online information sharing using the Grip op Water Altena web-platform is largely one directional and from authorities to citizens. Citizens also share information via this web-platform, but this has been so far been occasional and very limited in scope.

Although the realized outcomes of Grip op Water Altena are (yet) limited, creating a small community of stakeholders around the topic of pluvial flooding, awareness raising about participatory approaches for reducing the risk of pluvial flooding among both local community members and authorities, and creating new forms of communication and interaction between a small group of stakeholders were identified as the realized outcomes for this CBM. No major change was expected to take place in the near future because of Grip op Water Altena, but this CBM has the potential to contribute to more awareness raising, data sharing, and better communication and interaction among the stakeholders.

Evaluating the impacts of Grip op Water Altena is out of the scope of this research. The impacts of a CBM initiative take a long time to take place, i.e. based on Phillips et al. (2014), a period of 5-10 years. The impact of Grip op Water Altena is therefore (yet) largely unknown. Nevertheless, if the CBM continues and manages to engage a larger number of community members, it has the potential to improve communication about the topic of pluvial flooding among stakeholders, increase awareness about this issue, and contribute to trust building and transparency among stakeholders. Moreover, Grip op Water Altena has the potential to contribute to a change in attitude towards environmental stewardship and also help promote participatory approaches for collaborative management of water resources. Nevertheless, there is no guarantee that such positive impacts happen in the future. ANV has agreed to provide the organizational support for continuation of Grip op Water Altena and the authorities will provide the financial support for its activities in the future. Nevertheless, if in the near future, Grip op Water Altena fails to provide tangible benefits or fails to maintain or raise the interest of local stakeholders, the chances are high that this CBM and the required support for its continuation stops.

What constitutes 'community' in Grip op Water Altena?

Community in the context of Grip op Water Altena consists of a group of members who are interested in, concerned about or have stakes in the issue of pluvial flooding in Altena. In November 2019, the core members of this CBM include the Regional Water Authority Rivierenland, the Municipality of Altena, a few environmental NGOs and small number of local community members. Although there was an existing interest by the Regional Water Authority Rivierenland to work more closely with citizens, this community did not emerge on its own; rather, its initiation was supported by the Ground Truth 2.0 project. The vision, mission and objectives of this CBM were co-created in consultation with local stakeholders. The focus on the issue of pluvial flooding was already identified during the proposal stage and did not change during the co-design process. The number and composition of members in this CBM has fluctuated over time, but overall this initiative failed to maintain and increase the number of its members. Compared to the local community members, the Ground Truth 2.0 team and the authorities involved exerted more influence on shaping this CBM. Nevertheless, this community co-created new ways of online and offline interaction, communication and information exchange.

4.5 CONCLUSION

This chapter presented the findings of the baseline analysis of the Dutch case study, as well as the establishment process and results of Grip op Water Altena. This is in line with the second and third objectives of this research that are to test the empirical applicability of the conceptual framework and to evaluate the evolving processes, outputs and outcomes of CBMs over time. The following conclusions sum up the most important insights generated from studying this case. It should be clarified that the aim of presenting the following conclusions is not to generalize the findings of this particular case, rather it is to provide insights that may inform future studies or the establishment processes other CBMs. Therefore, generalizabilituy of the following conclusions is not suggested or implied by the author.

Alignment of CBM objectives with existing formal and informal structures and established ways or working of different actors is a challenge that requires a careful study of those structures and processes. Moreover, successful establishment of a CBM with the higher aim of supporting environmental stewardship and cooperative planning in a highly institutionalized context that does not envision or allow for public participation is difficult, if at all possible.

Mass participation in a CBM is not only determined by ease or difficulty of participation, but also by factors such as social and institutional contextual settings in which a CBM is being established, the need for its establishment (demand-driven versus supply-driven), as well as length of the establishment process. Social and institutional contextual settings such as the perceived importance and urgency of the topic, institutionalized role of different actors and the level of trust in authorities to manage the issue in focus of the CBM determine the number of participates in a CBM. Moreover, a lengthy process of establishment, during which less tangible results are produced, increases the risk of discouragement of participants and increases the number of drop outs. This is especially a challenge for co-designed CBMs that have to invest a lot of time on consensus building. Therefore, co-designed CBMs should set a clear timeframe for defining the aims, objectives and functionalities and communicate this with involved stakeholders at the very beginning to avoid future disappointments.

Although co-designing may provide a more equal chance for parties involved to influence the establishment processes of CBMs, this should not be misinterpreted as power relationships between different actors do not exist or are balanced out completely. In order to be able to understand the power dynamics in a CBM, existing power relationship among the actors, the issue of data ownership, source of technical, financial and organizational support for the CBM, and interests of actors involved in the establishment process should be carefully studied.

Technologies developed in Grip op Water Altena are accessible, user-friendly and barely exclude different stakeholder categories from their use. Nevertheless, the design of certain functionalities in this CBM (e.g. the decision for not having short communication lines) was influenced by the preference of authorities, which is an example of how vested interests of stakeholders with more authority and power can shape the functionalities of a CBM.

CBM initiatives can provide opportunities for interaction and information exchange; possibilities that have the potential to contribute to broader social, environmental and governance-related changes in the future. Nevertheless, outcomes and impacts of a CBM take time to materialize and become tangible or measurable. The timeline of this research did not allow for studying the medium and long-term changes resulting from Grip op Water Altena.

4. Results and discussion of the Dutch case study: Grip op Water Altena

98

5

RESULTS AND DISCUSSION OF THE KENYAN CASE STUDY: MAASAI MARA CITIZEN OBSERVATORY[24]

This chapter presents the results of the two phases of empirical research for evaluating the baseline situation of the Kenyan case study, as well as the establishment process and results of MMCO. The CPI Framework is used as a guiding frame for presenting the findings in this Chapter. This chapter presents a practical example of testing the empirical applicability of the CPI Framework for studying the evolving processes and results of a CBM, which is in line with objectives 2 and 3 of this research. Section 5.1 presents the background information of this case study and introduces MMCO to the reader. The results of the first phase of empirical research (i.e. baseline situation) in the Kenyan case study are captured in Section 5.2. Subsequently, the results of the second phase of empirical research (i.e. evaluation of the establishment process and results of MMCO) are presented in Section 5.3. Finally, the discussions and conclusions of this case study are presented in sections 5.4 and 5.5.

---

[24] This chapter is partially based on: Gharesifard, M., Wehn, U., & van der Zaag, P. (2019a). Context matters: a baseline analysis of contextual realities for two community-based monitoring initiatives of water and environment in Europe and Africa. *Journal of Hydrology*, 124144. doi:https://doi.org/10.1016/j.jhydrol.2019.124144

5.1 BACKGROUND OF THE KENYAN CASE STUDY

The Kenya case study is located in Narok County in southwestern Kenya, close to the Tanzanian border. This area includes the Maasai Mara National Reserve, the Mara Triangle and conservancies around the national reserve. This is part of the wider Mara ecosystem on the Kenyan side that is being managed by the Narok County government. Narok County has a total surface area of 17,921 km^2 and a population of 850,920[25] inhabitants. The majority of the inhabitants of this area are Maasai pastoralists and it is one of the most famous touristic destinations in Kenya. Human-wildlife conflict is a prominent issue in this area and practices such as overstocking, overgrazing and fencing alongside droughts have put a lot of pressure on biodiversity and people's livelihoods. Boles et al. (2019) argues that a post-colonial and tourism-led model of conservation has resulted in marginalization of local pastoralists and limited pasture for their livestock. At the same time, previous research has shown that existing regulations and decades of interventions did not yield a meaningful positive effect on limiting overgrazing and overstocking in the Mara (Homewood, 2004). Therefore, issues such as over stocking and overgrazing in the Mara can be considered "wicked" problems, for which there are no easy solutions.

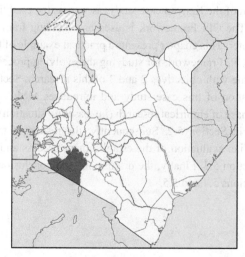

Figure 5.1 Location of the Narok County in Kenya
Source: Wikipedia[26]

---

[25] Based on the last official census in 2010; provided by the Kenya National Bureau of Statistics; retrieved from: https://www.knbs.or.ke/overview-of-census-2009/
[26] By Karte: NordNordWest, Lizenz: Creative Commons by-sa-3.0 de, CC BY-SA 3.0 de, https://commons.wikimedia.org/w/index.php?curid=38844143

The initial focus of the Kenyan case study of the Ground Truth 2.0 was on biodiversity conservation in the Mara. The ambition of this case was to establish a CBM that enables communication among local stakeholders and Masai Mara visitors and generates crowd-sourced biodiversity-related data and information (Ground Truth 2.0 consortium, 2015). Data and information generated in this CBM were expected to provide the Narok County Government, local community members, tourists and other stakeholders with better insights about the status of biodiversity in the Mara and help with better management of natural resources and sustainable tourism in this area.

IHE Delft Institute for Water Education, Upande and TAHMO were the core project partners involved in this case. The number and composition of the Ground Truth 2.0 team members involved in the establishment process of this CBM changed over time, but it normally included 7 or 8 people from the three aforementioned organizations.

The aims, objectives and functionalities of the CBM in this case were co-created in consultation with local stakeholders during 4 co-design workshops. The first co-design workshop was organized in Talek in March 2017, with 18 representatives of different organizations and local community members. Organizations represented were the Narok County Government, Kenya Wildlife Service, Kenya Meteorological Department, African Conservation Center, National Museums of Kenya, Egerton University, Maasai Mara Wildlife Conservancies Association (MMWCA), Kenya Wildlife Trust, and the Sand River WRUA. Moreover, a few local community members, one tour guide and a representative from Mara Loita Hotel were also present. Invitation for participation in the first co-design meeting was sent by Upande and based on their previous understanding of the stakeholders in the region from a previous project called MaMaSe (Mau Mara Serengeti)[27].

The agreed upon name for this CBM is Maasai Mara Citizen Observatory, or MMCO in short. The first co-design workshop focused on defining the problem and understanding the focus of the CBM. This meeting already showed a clear difference in interests of the representatives of the Narok County Government and locals community members. While the representatives of the county wanted to focus on the issue of biodiversity management, the local community members were interested in focusing on the issue of sustainable livelihood management for the local communities. In order to accommodate these two wishes, the central challenge of MMCO was defined as "balancing livelihoods and sustainable biodiversity management in the Mara ecosystem".

The next three co-design meetings were held between May 2017 and November 2019 in Narok and focused on co-creating the functionalities of MMCO. The outputs of these

---

[27] http://www.mamase.org/

meetings resulted in developing a web-platform[28], as well as the Mara Collect and the MMCO Apps. The Mara Collect App uses ODK; an open-source software that allows for collection of data without internet connectivity. The Mara Collect App allows for collecting data about emergencies, incidents, biodiversity, scenery, pollution and natural hazards. The MMCO App and platform are designed for sharing the collected information using the Mara Collect App. Local community members, researchers, rangers, local authorities, tourists and tour guides are among the expected end-users of these tools. The data collected and shared using these technological components was expected to help the local communities and the authorities in a variety of ways. For example, provide local communities with a better overview of available water resources and grazing lands and help the authorities to predict species' distributions. Some of the aforementioned data (e.g. location of endangered species) were considered sensitive and agreeing on a data sharing policy became a lengthy process. In addition to the data and information submitted via the Mara Collect App, the MMCO App and the web-platform incorporate the data collected by a number of in-situ sensors, such as weather stations and water level sensors. These stations were either installed as a part of the Ground Truth 2.0 by TAHMO[29], or were being maintained by Upande or TAHMO from previous projects. The data from these sensors was not considered sensitive and became available on the MMCO App and the web-platform almost immediately. Some screen shots from the two Apps and web-platform are presented in Annex 7.

The first version of the web-platform and the two Apps was first launched during the March of the Elephants Day in the Mara in September 2017. Subsequently, an updated version of the web-platform and the two Apps was presented during a public event at the World Wildlife Day in March 2018. These were called soft-launches due to the fact that a data sharing policy for this CBM was not yet agreed upon.

Using the Apps required a short training and therefore two training sessions were organized inTalek and Oloisukut conservancy, during which 70 community members, KWS rangers and conservancy rangers were trained in using the MMCO Apps. In addition, two other face-to-face meetings were organized in August and November 2019, in which MMCO members discussed its sustainability. Both of these meetings were hosted by the Maasai Mara University. The November 2019 meeting also marked the official launch of the two Apps.

The participants in the aforementioned face-to-face meetings and those who use the tools developed in this CBM online are considered members of MMCO. For planning purposes, almost all participants in MMCO face-to-face meetings were invited to attend these meeting, nevertheless occasional spontaneous participants were also among the attendees.

---

[28] https://mara.info.ke/
[29] https://tahmo.org/

Invited participants in the meetings received compensation for their travel costs and were provided with meals and accommodation (in case they were travelling from a distance). Using the MMCO Apps and web-platform was free and did not require registration.

Recruiting members in MMCO proved to be difficult and the Apps and the web-platform of this CBM have been so far only used to a limited extent and mainly in training sessions. This is partly because of the context-related factors such as complexity and sensitivity of the issue that the CBM is engaged with, but also because of internal factors such as the design of the tools in this CBM. The baseline situation in which MMCO was established and factors that affected its establishment, functioning and results are discussed in detail in sections 5.2 and 5.3.

5.2 BASELINE SITUATION OF THE KENYAN CASE STUDY

Similar to the Dutch case study, the CPI Framework was used as a guiding frame for analyzing the baseline situation of the Kenyan case study. In order to do so, context-related aspects of 'power dynamics', 'participation' and 'technology' dimensions of the CPI Framework were carefully studied (Figure 4.3). This section presents the findings of this baseline analysis.

5.2.1 Power dynamics in the Kenyan case study

Social, institutional & political context

Given the fundamental changes in the Constitution of Kenya (CoK) in August 2010, the country's institutional framework took a major turn from a centralized system of administration and governance to a devolved structure of government which cedes power of legislation, execution and also revenue collection and expenditure to 47 counties. The judiciary power, however, remained central at the national level. With these changes, the new constitution introduced two main levels of government: national and county. In terms of hierarchy, the constitution defines these two levels as parallel and does not indicate any superiority for one level over the other (Government of Kenya, 2010- Article 1), but several Articles emphasize the need for consultation collaboration and coordination between the two levels. For example, Article 6 (2) defines these levels as 'distinct and inter-dependent', but highlights the need for their 'mutual relations on the basis of consultation and cooperation'. Two main points were elicited from the interviews. Firstly, the county is autonomous, but its decisions should not conflict with the national (parliament) law and in any conflicting case, the national law prevails. Secondly, a number of interviewees believed that according to the devolution of power, the county government has the final decision making power; yet in many cases it is still the national government that takes some of the measures on the ground.

103

Apart from the CoK, the interviewees identified (inter)national-level legislation and policy guidelines related to the topic of the CBM, including the agreement for the joint trans-boundary management of the Mara River Basin, the Kenyan law on the management of protected areas, the Wildlife Act, the Water Act 2005, and the Forest Act. At the county level, the Narok County Integrated Development Plan and the Environmental Management Bill are the main pieces of legislation and, for many other aspects, the county uses bylaws driven by the national policies and customized to the local issue. Nevertheless, at the time of conducting this research, the Narok County Government was in the process of approving a number of bills that will substitute these bylaws, namely the Maasai Mara Management Plan, the Narok County Tourism Act, the Environmental Management Act and the Livestock Act.

Implementation of the aforementioned rules and regulations often involves informal, complex and undocumented decision making processes that are very difficult to study. The vast majority of the interviewees believed that the extent of implementation of rules, roles and responsibilities regarding biodiversity conservation and livelihood management in Kenya highly depends on the specific issue, location and actors involved, but, overall, implementation is not very strict at both local and national levels. Interviewees often referred to increasing wildlife deterioration and the existence of multiple barriers for sustaining livelihoods to back up their claim. They identified several institutional, social and political barriers for improved implementation, including lack of resources (money, knowledge and staff); especially at the county level, conflicting personal and political interests and corruption. Moreover, it was mentioned that strengthening implementation by the county is difficult, because the county does not have the judiciary power to enforce the law; for example, in a poaching incident, only the Kenya Wildlife Service (KWS) can arrest and prosecute people; if the county rangers want to arrest someone, they have to report it to KWS. The same applies to the issue of illegal logging and the Kenya Forest Service (KFS).

Interviewees also highlighted several values, norms and traditions of the Maasai culture that can positively or negatively influence biodiversity conservation and livelihood management in the Mara. For example, the traditional land ownership model of the Maasai was community land ownership instead of individual land ownership. Preserving this tradition can help stop controversial practices such as fencing, which increasingly cause wildlife deaths. On the other hand, Maasai are traditionally pastoralists, and they consider the number of livestock they have as an indicator of their wealth; therefore, overstocking and overgrazing have became prominent issues in the Mara region.

Baseline situation of authority and power of stakeholders

All interviewees from the regulatory entities at the county and national levels perceived to have direct influence on decisions related to biodiversity conservation and sustainable livelihood management because of their official mandate. According to the CoK and a number of interviewees from the local regulatory entities, the county government has the strongest influence among all actors. Nevertheless, local chiefs play a vital role in the balance of power between the national and local government. They have an explicit influence on daily local decisions and act as a direct link between the community and national authorities. They have an advising/consulting role towards the national government and relay the decisions and agreements that are made at the national level to the community. Moreover, the church was also identified as one of the most powerful stakeholders; a platform which can communicate directly with the authorities and influence public opinion. In this regard, a spiritual leader from a church said: "we provide advice and consultation to the authorities; even the president sometimes consults the church on different decisions"[30].

The interviewees from the CBM co-design group, the expert advisors and the general public expressed two ways in which they can have a (minimal) influence on decisions: (1) via altering public opinion (rather than policies and decisions) and educating community members (e.g. via awareness raising programs); or (2) by providing advice to the authorities inside or outside of participatory meetings. The rest of the interviewees from these groups stated that they have little or no influence on decisions and they are informed about decisions when these are gazetted or advertised via other channels. In this regard, one interviewee mentioned: "our chance to influence the decisions is very limited; they [the authorities] are not ready to listen to just a simple person"[31].

Some of the interviewees from the general public also emphasized that at household level, the decision making power normally rests with men; female members of the communities are less involved in making decisions.

Baseline situation of access to and control over data and information

Article 35 of the CoK (2010) explicitly indicates that every Kenyan citizen has the right of access to the information held by the State, or by another person, if this is required for exercising or protecting any right or fundamental freedom. Furthermore, the state shall publish and publicize any important information affecting the nation. With reference to this Article, the Access to Information Act (2016) and Part IX of the County Government Act (2012) preserve the right of Kenyan citizens for requesting access to information held by any government organization, including the county government or any unit or department thereof. Despite these written rules, a great majority of the interviewees believed that the level of access to data actually depends on the organization, department or individuals who hold the data. Several interviewees criticized the current (limited) data

---

[30] KE-01-23
[31] KE-01-27

sharing practices of the authorities, based on their previous experiences and encounters. For example, an interviewees mentioned "sometimes they give you the data, and sometimes they simply decide not to"[32]; another one argued that "sometimes you will be asked why you need the data in a harsh way; they have the fear of use of this information by the community"[33]. This is due to the fact that certain data and information are deemed 'sensitive' or 'exempt information' and although defined in the legislation, the definition is open to interpretation. This has resulted in keeping essential information from the public. Moreover, some expert advisors stated that, in any case, one should not publish data that is not accepted or endorsed by the government; otherwise they will advertise against using the data and claim that the data is not good at all.

All interviewees unanimously believed that the availability of data about biodiversity and livelihoods is not very good. The available data was often described as inaccurate, unreliable, not up-to-date, not well distributed (spatially and temporally), and inaccessible; especially publicly or in a central way. Furthermore, lack of coordination between organizations in terms of collecting and sharing data was identified as another issue. It was mentioned that if data exists, it is difficult to know who has the data and the organizations who hold the data do not share it with the public and sometimes not even with each other.

Research organizations, individual experts/scientists, NGOs and government organizations were identified by the interviewees as the best-placed stakeholders for analyzing data on biodiversity and livelihoods; however, it was mentioned that for some data, the analysis requires highly specific expertise that may not even exist among the county rangers. On the other hand, many interviewees believed that the local people have a very good understanding of biodiversity and livelihoods, but literacy and language are barriers that affect their ability to share their knowledge.

---

[32] KE-01-20
[33] KE-01-21

5.2.2 Participation dynamics in the Kenyan case study

Baseline situation of participation in policy and decision making processes

The Parliament of Kenya was identified as the main body of policy making regarding biodiversity conservation and livelihood management at the national level. Cabinet secretaries, especially those who head the ministries, involved in managing biodiversity and livelihoods, relevant wings of ministries such as KWS, KFS, Kenya Water Resources Management Authority (WRMA), the National Museums of Kenya and the Kenya Water Tower Agency were identified as the most relevant actors for the issue in focus of the CBM initiative at the national level. Furthermore, the Narok County Government was elicited as the most influential stakeholder at the local level. Local chiefs, conservancies and NGOs were also deemed relevant stakeholders at the local level. Despite the devolution of power from the national level to the county government in Kenya in 2010, the national level organizations such as KWS, are still very much involved in conservation activities at the county level.

Public participation has a vivid footprint throughout the CoK (2010) and in several other pieces of legislation including the County Government Act (2012), the Public Service (Values and Principles) Act (2015), and more specifically related to the topic of this CBM initiative, the Environmental management and Coordination Act (1999), the Wildlife Conservation and Management Act (2013) and the Agriculture and Food Authority Act (2013). The first Article of the constitution portrays public participation as a right for every Kenyan and states that all sovereign power belongs to the people of Kenya and that they can exercise their power directly or through their elected representatives. The Constitution also recognizes participation of the Kenyan people as a national value and a governance principle and obliges the state to facilitate participation of everyone in governance processes. Furthermore, public participation is also mentioned as a purpose of devolution of government for providing self-governance power to the people and to enhance their participation in decision making processes. The national and county governments are obliged to facilitate and encourage public participation in the legislative process, public finance and also conservation of environmental and natural resources management.

Despite the formal anchoring of participation in these legal frameworks, none of the interviewees perceived any involvement of the general public in policy and decision making processes regarding biodiversity conservation and livelihood management. This, among other things, presents a picture of the limited extent of implementation of these legislations. Despite much emphasis on public participation in the CoK and the aforementioned acts, Kenya still lacks a national act that provides a general framework for how to give effect to the constitutional and legislative requirements in this regard. It has been argued that such a framework at the national level is needed to guide the public participation processes at the local level (Marine, 2015). The Public Participation Bill (2016) was prepared to cover this gap; however, at the time of conducting this research, this bill still has not been accepted due to controversies and opposing ideas and the County Public Participation Guidelines (2016) was the only available (but not legally binding) tool to guide this process.

Existing patterns of communication

There is no article in the CoK or other legislation that directly mentions the desired pattern of information flow between different stakeholders for policy and decision making processes; however, several articles portray a 'unidirectional' communication paradigm as the prescribed pattern of communication from state and county government to the general public. The main purpose seems to be informing the public about decisions. As an example, the County Government Act (2012) requests county governments to integrate communication in all development activities.

Using telephone and smart phone App calls (predominantly WhatsApp) are by far the two most frequently used means of communication in this case study. Due to the low average level of literacy among local citizens, phone calls are much more common than sending SMS. Different communication channels are used for reaching different target groups; for example, social media (e.g. Facebook) is one of the best channels for reaching out to the youth, while for elders face-to-face communication works better. Moreover, radio was indicated as an efficient channel for reaching a large number of people, even in remote areas. The majority of the interviewees mentioned that they communicate about biodiversity conservation and livelihood management both for work and personal purposes. The authorities mainly prefer to communicate with community members face-to-face, while the community members mostly use face-to-face meetings, phone calls and SMS, WhatsApp and email for reaching out to the authorities. Figure 5.3 presents the preferred channels for communication. As the figure shows, using an App on a smartphone (predominantly WhatsApp) and using websites or blogs are, respectively, the most and least preferred channels for communicating about biodiversity conservation and livelihood management in this case study.

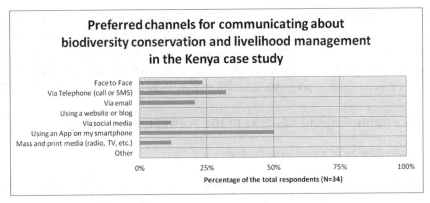

Figure 5.2 *Preferred channels for communicating about biodiversity conservation and livelihood management in the Kenya case study*
Source: Gharesifard et al. (2019a)

Existing methods of communication and participation in decision making processes

In the Kenyan case study, all participants from the regulatory entities perceived they have a direct or indirect say in the decisions because of their technical expertise. The expert advisors indicated that they are mostly involved via their jobs and the nature of their involvement was highly dependent on their work. This level of involvement ranged from expressing opinions by communicating about their research, to deliberation and negotiation with the authorities (see Fung, 2006).

Many interviewees from the CBM co-design group and the general public perceived that they are barely or not at all part of decision making processes related to biodiversity conservation and livelihood management. Some of these interviewees indicated they only get to know about decisions when these are gazetted. The other interviewees from the CBM co-design group and the general public identified two main methods by which they are involved in decision making processes. The first method was expressing preferences in participatory meetings or via membership in community action groups. Although it is not always possible for everyone or even the community representatives to participate in such meetings (because of the location or timing of the meetings), those who do participate have the opportunity to express their preferences and even bargain with the authorities. For example, a local Maasai gave a clear example of bargaining power that he had because of his membership in a local water users association: "When we participate in these meetings, we express our opinion and contribute to the discussions; for example, through appealing to the 2010 constitution and the recent Wildlife Act we could get them [the Narok County Government] to assign more compensation for livestock deaths"[34]. The second identified method was by communicating with different (influential) stakeholders bilaterally outside of participatory meetings. Interviewees highlighted that using this method depends heavily on social or professional links with these stakeholders. In this regard, one interviewee mentioned "if you want to have a good influence, you should be present in the political forums, or know people who are there"[35].

5.2.3 Technological context in the Kenyan case study

Baseline situation of access to technology

Mobile phones are the main medium for digital communication in Kenya and at the time of conducting this research there were 81.3 postpaid and 'active' prepaid mobile-cellular telephone subscriptions per 100 inhabitants in Kenya (ITU, 2017b). In contrast, only 0.15% of residents had fixed-telephone subscriptions, less than 15% of households had computers and 22.3% of households had internet access.

Although a large proportion of the population used mobile phones, only some of these individuals were (actively) using the internet (World Economic Forum, 2016a). The average low level of literacy in Kenya is a challenge for developing the required skills for effective use of ICTs. In 2017, the 'mean years of schooling' in this country was only 6.3 years.

In terms of both access to and the skills to use ICTs, Narok County was performing below the national level. The strategy for spatial development of ICT infrastructure in Kenya is very much focused on service provision for population centers. For example, in 2016, with a geographical coverage of only 17%, up to 78% of the Kenyans had access to 3G broadband service (Intelecon, 2016). The downside of this strategy is that a large part of the counties, including a large proportion of the Narok County that has a dispersed population, are either underserved (e.g. do not have access to 3G broadband) or not served at all. Moreover, a report by the Kenya National Bureau of Statistics indicates that in 2013 only 11% of Narok County inhabitants had a secondary level of education and 38% of the residents in this county had no formal education (Ngugi et al., 2013).

Baseline situation of included/excluded groups from use of technology

Despite advancements in the ICT sector in Kenya, there is still a big gap in both, access and skills to effectively use technology. In addition to differences in accessing technology, the low level of literacy (especially among the older generation) and gender discrepancies in access to and use of technology are contributing factors in creating new

---

[34] KE-01-03
[35] KE-01-26

forms of inequalities (Brännström, 2012; World Wide Web Foundation, 2015). The gap in access and use is more significant in rural areas with dispersed population and can highly affect the ability of community members to take part in public participation processes. As an example, public participation in the process of formulating a national strategy for wildlife conservation and management in 2017 was only possible by sending a written memorandum via email or expressing oral or written opinions in the 'structured regional meetings'. Sending a written memorandum via email, among other things, requires a certain level of literacy, computer skills and internet access, provided that community members were aware of this consultation in the first place. Moreover, structured meetings at the regional level are normally held at the center of each county, which also limits the chance of participation for community members who live in remote areas.

5.3 THE ESTABLISHMENT PROCESS AND RESULTS OF MMCO

This section depicts the establishment process and results of MMCO. Sub-sections 5.3.1 to 5.3.5 correspond with the five dimensions of the CPI Framework and help answer research questions 1 to 5 of this study. Similar to the Dutch case study, the results reported in this section are informed by phase 2 of empirical research and reflect the status of MMCO at the end of November 2019 (i.e. the end date of data collection in phase 2 of the empirical research).

5.3.1 Objectives and actor specific goals in MMCO

Based on an evaluation of the overarching objectives and actor-specific goals in MMCO, this section provides the answer to the first research question for this CBM, i.e. what are the overarching objectives and actor-specific goals of MMCO and to what extent does the design of this initiative help achieve those goals/objectives?

Overarching objectives of MMCO

During the initial phases of the co-design process of MMCO, different stakeholders agreed on a vision and mission, as well as four specific objectives for this CBM, which are presented in Table 5.1.

Table 5.1 Vision, Mission and Objectives of MMCO

Vision
"We envisage a society in which all stakeholders are working together to ensure the balance between sustainable livelihoods and biodiversity management in the Mara ecosystem"
Mission
"The citizen observatory will constitute a multi-stakeholder platform for generating and sharing of data, information and knowledge to improve policy making and implementation for sustainable livelihoods and biodiversity management in the Mara ecosystem"
Objectives
Obj1. "To provide a monitoring system for biodiversity, livestock and crop, land and water resources, and climate across the Mara ecosystem by 2017"
Obj2. "To establish a repository on Mara biodiversity, livestock and crop, land and water resources, and climate information that is accessible to all stakeholders by the end of 2017"
Obj3. "To develop a platform by the end of 2018 for the engagement of citizens, government, research and the private sector to promote practices that create the balance between livelihoods and biodiversity in the Mara ecosystem"
Obj4. "To improve data, information and knowledge generation and sharing on biodiversity and livelihoods between citizens, practitioners, researchers and policy makers by 2018 for informed policies and policy implementation"

Interviewees from both the GT2.0 team and members of MMCO were asked to reflect on these objectives by (1) indicating the most important objective(s) in their view, (2) reflecting on the extent of achievement of the objectives, and (3) (if any) identifying shortcomings in achieving the objectives.

The most important objective of MMCO for the GT2.0 team members was to create a CBM initiative, through which data and information is generated and shared with the community members and the authorities. The team members described this as a socio-technical platform, which consists of people and technology. Increased access to data and information because of this CBM was expected to support better decision and policy making, which in turn would lead to a better balance between biodiversity conservation and livelihood management.

According to the majority of the CBM members, improving data, information and knowledge generation and sharing on biodiversity and livelihood management was the most important objective of MMCO. Most of the answers implicitly referred to establishment of a monitoring system for achieving this objective, which is closely linked to the third objective of MMCO. Three interviewees believed that creating a platform for bringing stakeholders together and engaging the communities, especially those who are living in the areas close to the wildlife and natural resources is a very important objective of this initiative. Moreover, two interviewees mentioned creating a repository of information e.g. on market prices of livestock, floods and rainfall is a very important objective of MMCO, because it helps addressing local issues and making decisions both by the community members, as well as the authorities.

All GT2.0 team members and the majority of the interviewees from the CBM members believed that the objectives of MMCO were partly achieved. The only exceptions were one of the MMCO members who believed that the objectives are completely achieved and two interviewees who were not sure about the extent of achievement of the CBM objectives.

Interviewees identified multiple reasons for partial achievement of the objectives. Engaging 'the right people' proved to be difficult for a number of reasons and this affected achieving the objectives of this CBM initiative. Some organizations (incl. Maasai Mara Wildlife Conservancies Association and the Narok County Government) sent junior staff members to the meetings and the lack of internal communication at those organizations hindered conveying the message across to the higher levels of management. There was also a lack of interest and a low sense of local ownership among some key stakeholders. This was partly due to the fact that this was an EU-funded project-driven CBM that focused on very sensitive and contested local issues, and even a Nairobi-based partner of the project (i.e. Upande) was perceived as an outsider; a tech company that is looking to make a profit.

The GT2.0 team members believed that another major factor that resulted in a low level of engagement and actual use of the App and web-platform was the issue of data sharing policy. A part of the data and information that was planned to be collected and shared using the Apps and the web-platform was deemed sensitive by some stakeholders (e.g. the Narok County Government) and this called for having a data sharing policy. The need for this data sharing policy was realized half way through the project and the participatory process of drafting, discussing and agreeing on its content took much longer than expected (more specifically until November 2019). While the tools for collecting data were in place, they could not be rolled out at a larger scale because the data policy was not ready and this negatively affected the momentum in the project and resulted in a low level of engagement of the end users. Interestingly, no CBM member elicited the data policy as a hindering factor for engagement; instead, they criticized the low number of community representatives in the collaborative meetings and mentioned that this negatively affected the dissemination of the results across community members.

Cultural norms were also identified as contributing factors that affected the number of people who could be invited to the face-to-face meetings. As a widely practiced norm in majority of projects in the study area, local stakeholders expected to receive a Daily Subsistence Allowance (DSA) for participation in face-to-face meetings of projects. The cost implications of this norm is one of the issues that affected the number of people who could possibly be invited to the meetings, and will remain an issue if there is a need for face-to-face interaction among the stakeholders after the end of the Ground Truth 2.0 project. Moreover, issues such as low level of literacy, access to technology and internet, and low interest in digital technologies among the local communities (especially the elderly) were identified as contributing factors to the low level of participation in using the CBM tools.

Goals of different actors in MMCO

For any initiative that aims to bring a large number of stakeholders together, coming across diverse goals, interests and wishes is not unusual, and in fact is to be expected. A difference in interests of local community members and authorities in this case resulted in defining the central challenge of MMCO as "balancing livelihoods and sustainable biodiversity management in the Mara ecosystem". Having a central challenge with a very wide focus had both positive and negative implications that are discussed hereafter.

When asked about the reasons for their participation in MMCO, several CBM members mentioned that they find the topic or activities of this initiative relevant for their professional purposes or personal interests and therefore they had an interest in the generated and shared data and information. For example, organizations such as the Narok County Government, MMU, KWS, KFS and conservancies found the topics in focus of the CBM relevant for their activities and mandates. Moreover, biodiversity and livelihood-related data that was being generated and shared in MMCO was deemed valuable by different organizations and to the local communities, simply because it was considered as new data and information that did not exist before or was not (easily) accessible to them and included a wide range of attributes that could be used for different purposes. Given the wide focus of MMCO, which can satisfy various data needs, all stakeholders could find a potential benefit for their organization, themselves or their community.

Nevertheless, defining a very wide central challenge and focusing on a large number of observations also had a downside. Stakeholders who had an interest in specific information found the data produced in MMCO dispersed and very generic. For example a high level county official believed that "the information [produced in MMCO] should have been customized based on the needs of different organizations. Having general information is good, but not sufficient; the outputs should have been divided based on the needs of different stakeholder groups"[36]. Furthermore, trying to be inclusive and appealing to the many wishes of the stakeholders created a sense of mistrust in the quality of the data produced, because stakeholders were uncertain about which organization has the capacity and expertise to be in charge of quality control of the data.

In addition, a specific conflict of interest was identified that relates to a parallel effort for developing an App for monitoring biodiversity. This App is called WILD (Wildlife Information Landscape Database) and aims at supporting data collection and sharing about biodiversity-related issues such as animal mortality, poaching, human wildlife conflict and illegal human activities. This App was developed by @iLabAfrica (Strathmore University) in partnership with USAID, and it was launched in September 2016[37]. Maasai Mara Wildlife Conservancies Association (MMWCA) was involved in the development process and rolling out of WILD, which included training 104 rangers and 11 conservancy managers for using the App[38]. MMWCA's involvement in this parallel initiative resulted in less interest and commitment from their side to get involved in MMCO.

Monitoring of the objectives in MMCO

There was no formal procedure specifically designed for monitoring the objectives of MMCO; however the GT2.0 team members identified two mechanisms that helped with monitoring and reflecting back on achievement of the MMCO objectives.

The first mechanism was revisiting the objectives during the collaborative meetings. From the moment that the MMCO objectives were co-created and agreed upon, these objectives were revisited regularly in almost all face-to-face meetings with the stakeholders. This provided all stakeholders with an opportunity to revisit and reflect back on the objectives.

Similar to the Dutch case, since May 2018 'reverse impact journey' (or reverse objectives journey) was used by the Ground Truth 2.0 team to reverse engineer the activities needed for achieving the co-created objectives of MMCO. For example, the forth objective of MMCO, which is improving data, information and knowledge generation and sharing on biodiversity and livelihoods, is only achievable through commitment of different actors for collecting and sharing this data and information. This means that there was a need for

---

[36] KE-02-19
[37] https://ilabafricastrathmore.wordpress.com/2016/09/20/the-wild-app-helping-conserve-our-wild-animals/
[38] https://www.maraconservancies.org/tag/wildlife/

identifying and assigning roles and responsibilities to different actors regarding data and information generation and exchange.

Change of MMCO objectives over time

The MMCO objectives were defined using a co-design approach, with inputs from all participating stakeholders, and after a consensus making process. These objectives were defined quite broadly and there were no changes in the agreed-upon objectives of MMCO after this consensus making process.

5.3.2 Participation dynamics in MMCO

This section unpacks the participation dynamics in MMCO, and aims at answering the second question of this research; who participates in the CBM initiative and how, and who does not?

Type of initiative

Similar to Grip op Water Altena, the typology of CBM initiatives proposed by Wehn et al. (2015a) was used as basis to analyze the type of initiative for MMCO[39].

MMCO envisions "a society in which all stakeholders are working together to ensure the balance between sustainable livelihoods and biodiversity management in the Mara ecosystem". This vision refers to an Environmental Stewardship model, in which all stakeholders collaborate and share responsibilities for achieving a balance between sustainable livelihoods and biodiversity management in the Mara ecosystem. Based on the mission of MMCO, this initiative will be done via establishing "a multi-stakeholder platform for generating and sharing of data, information and knowledge to improve policy making and implementation for sustainable livelihoods and biodiversity management in the Mara ecosystem". This mission indicates an Environmental Monitoring model as defined by Wehn et al. (2015a); a CBM that aims at data collection and sharing about the topics of biodiversity conservation and livelihood management in the Mara. Therefore, the domain of this initiative is both environmental monitoring and environmental stewardship.

Similar to the Dutch case study, authorities such as the Narok County Government, KWS and KFS are officially mandated with the management of the issues in focus of this CBM. However, the institutionalized place of public participation in Constitution of Kenya and several pieces of legislation provide a legal basis for participation of all stakeholders in such processes, and if needed this can be leveraged as a right for participation.

---

[39] For further explanation about the typologies see section 4.3.1.

Geographic scope of MMCO

The term used to refer to the geographic scope of MMCO in its vision, mission and objectives is 'Mara ecosystem'. This is not limited to, but includes the Mara Triangle, the Maasai Mara National Reserve, and the conservancies around this reserve.

The state of the Mara Triangle, the Maasai Mara National Reserve and conservancies in the Mara is constantly changing and this includes the surface area of the land under conservation, number of hotels and lodges, as well as the number of landowner and rangers in the park and different conservancies. There is no up-to-date official source of information about the number of hotels and lodges in the Maasai Mara National Reserve, but some websites[40&41] list up to 25 hotels, lodges and camps inside the Maasai Mara National Park. Figure 5.3 presents a map of the Maasai Mara National Reserve, Mara Triangle and the conservancies around the reserve, as well as the most recent statistics about the state of the conservancies. This map was published in July 2019 by MMWCA i.e. the umbrella organization of the conservancies in the Mara region. Based on the information on this map, 12350 land owners, 274 rangers and 51 lodges in the conservancies were among the potential pool of participants in MMCO.

---

[40] http://www.zakenya.com/travel-leisure/a-list-of-hotels-in-maasai-mara.html

[41] http://maps.mamase.org

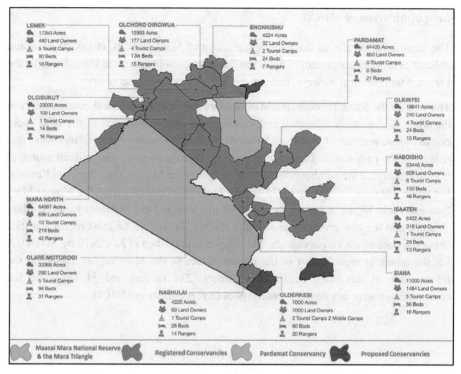

Figure 5.3 State of the conservancies in Mara region
Source: modified from Muli and Mbelati (2019)

(Non) Participant groups in MMCO

During the baseline analysis, a number of stakeholders that were relevant for MMCO's aims and activities were identified. The result of this analysis was presented in section 5.2.2 above. The aim of this section is to provide an overview of the stakeholder groups that in practice have participated in the process of designing the functionalities of MMCO, as well as the end users of its tools.

Representatives of several local and national level organizations were present in the co-design workshops, in which the functionalities of MMCO were defined. This includes representatives of different departments of the Narok County Government, KWS, KFS, Kenya Meteorological Department, National Museums of Kenya, and WRMA. Moreover, Egerton University, Maasai Mara University, and NGOs such as the African Conservation Centre, Friends of Maasai Mara and MMWCA were also represented. In addition, a number of individual community members who often represented organized community groups such as WRUAs and conservancies also participated in the co-design process.

Although the Narok County Government always participated in the co-design workshops, their representation was often at the junior level and lack of internal communication between the representatives and higher levels of decision making at the county affected the government's engagement with the project. Moreover, due to the fact that community members live in geographically dispersed locations, local pastoralists, chiefs and the private tourism sector (e.g. lodges and hotels) had limited representation in the meetings; an issue that affected the broader uptake of the initiative by local community members. Moreover, some important national level stakeholders such as the Ministry of Environment and Forestry, Ministry of Agriculture, Livestock, Fisheries and Irrigation, and Kenya Market Trust were not represented in the co-design meetings.

At the time of conducting this research, the uptake of the tools in MMCO was very limited. The weather stations had the most uptake, because they were being used by the universities and schools for education and research purposes. Moreover, the produced weather data was being used by the Meteorological Department of the Narok County for producing forecasts and alerts. Although the MMCO Apps have been downloaded 100+ times (which means between 101 and 500 downloads), the actual use of the Apps was very limited and the submissions were geographically sparse and occasional (See Figures 5.6 and 5.7). Indicating a name while submitting observations using the Mara Collect App is optional and users may use fake names to submit the data. Therefore, it is not possible to analyze which stakeholders have been most active using the App. However, the observations made during the fieldwork in November 2019 confirmed that main stakeholders such as the Narok County Government, KWS and KFS, were not (yet) using the MMCO App.

Google Analytics was used for analyzing the number of active users of MMCO Web-platform. In this analysis, unique users who visited the web-platform at least once within a 28-day period were assumed as active users. Figure 5.4 illustrates the results of this analysis that was done for a one year time period from 1st of December 2018 until end of November 2019. As the figure shows, there are no records of use of the web-platform before September 2019. This is because the MMCO web-platform was not registered for Google Analytics before September 2019. Nevertheless, if we consider the available data as an indication of the number of active users of MMCO web-platform, apparently only a few users (i.e. not more than 24) regularly visit this web-platform. This confirms one of the findings of the baseline analysis in this case that identified websites as the least preferred channel for communicating about the topics in focus of this CBM.

119

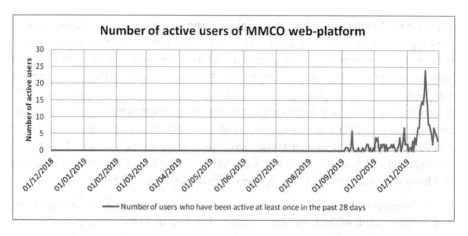

Figure 5.4 Number of active users of MMCO web-platform

Efforts required to participate in MMCO

Efforts required to participate in MMCO refers to the requirements and investments that are needed from the participants' side, in order to be able to use the MMCO tools or attend the participatory meetings of this CBM initiative.

The need to have access to a smart phone, the time people need to spend for data collection and sharing, and a number of financial implications for participation were among the most frequently mentioned responses. The need to pay for data bundles, commute costs (e.g. fuel) and costs of participation in meetings (e.g. accommodation and food) were among the identified financial burdens. For example, a KWS officer mentioned that if a ranger wants to respond to a reported human-wildlife conflict, he needs to use data bundle and may require extra fuel to travel to more places than his usual posts, resources that are often not compensated or may not be readily available[42].

Several interviewees from both groups also mentioned that using the MMCO tools requires a short training. This training can be through participation in training sessions, learning from other who have previously used the tools, or by using available manuals. A number of training sessions were organized for using the MMCO tools that are explained in the following section. Available manuals were limited to instructions provided by TAHMO using the weather stations.

---

[42] KE-02-20

Support offered for participation in MMCO

Support offered by MMCO can be divided into four broad categories; the technical support for development of the tools, financial support, organizational support, and the awareness raising and capacity building support that was provided by the Ground Truth 2.0 partner organizations.

MMCO tools include the two Apps, the websites, physical sensors (weather stations and water level sensors) and four screens that are installed for displaying the data in different locations. One of these screens is installed in the KWS office in Narok, two in conservancies (i.e. the Mara Triangle conservancy and the Oloisukut conservancy), and one at the G&G Hotel in Talek.

Organizational support include scheduling, setting up, hosting and moderating the participatory meetings and outreach events, as well as setting up and maintaining social media accounts. The costs of organizational support were covered by the Ground Truth 2.0 project, but from 2020 these costs should be covered by the revenue streams of this CBM, which by end of November 2019 were largely unknown (see section 5.3.3).

The fourth category relates to capacity building for using the MMCO tools and raising awareness especially within the communities about how they can contribute to conservation through making, sharing and using environmental observations. This includes training of participants in the meetings, training of trainers, and the available manuals for using the tools. The MMCO website provided a link to a number of manuals and teaching materials from the TAHMO School 2 School Initiative, which is mainly meant to be used by teachers to familiarize students with weather data. In addition, a manual for participation in mapathons was shared with the Maasai Mara University and a number of lecturers were trained on how to conduct mapathons, however this manual was not available on the MMCO website. No manual was available for using the MMCO app. Face-to-face trainings included the instructions sessions during the co-design workshops, as well as two dedicated training sessions that were organized for using the MMCO tools. Furthermore, a few staff members of the Maasai Mara University were trained on conducting mapathons. During the lifetime of the Ground Truth 2.0 project, the costs of holding the co-design meetings and training, including the costs of accommodation, transport and food for the participants, was covered by the project.

Despite of the aforementioned efforts, a number of interviewees believed that so far only a small percentage of the potential end-users of MMCO tools have been present in the meetings and trainings, or have heard about this initiative. Many more people still need a formal introduction and training to be able to join MMCO. However, at the time of conducting this research, the financial resources and the technical support required for these dissemination and outreach activities were not clearly defined and agreed upon by the MMCO members.

Pattern of communication in MMCO

The baseline of preferred communication channels in this case was presented in section 5.2.2. Using smart phone Apps (predominantly WhatsApp), telephone calls, face-to-face communications and using emails were the most frequently preferred communication channels (see Figure 5.2).

In the second phase of the empirical research, interviewees were asked to identify the communication channels, which they have actually used to participate in different activities of MMCO, and the extent of use of each channel. Figure 5.5 summarizes the results of this inquiry. Overall, the actual use of communication channels in MMCO shows a very good match with the preferred channels of communication identified in the baseline analysis and shows that the 'existing norms and mental frameworks for communication' (Gharesifard et al., 2019b) were followed by the members of MMCO to communicate with each other.

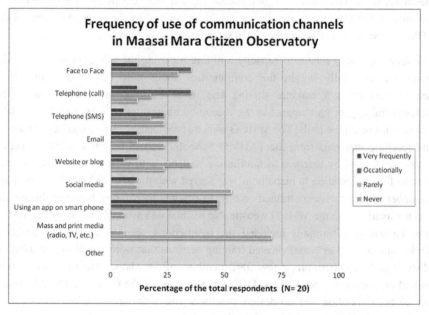

Figure 5.5 Frequency of use of different communication channels in MMCO

Interviewees were also asked to indicate for what purpose they have used each communication channel. The most frequently used channel was using an App on smart phone, but this mainly referred to the CBM WhatsApp group (rather than the CBM Apps). The MMCO members and the GT2.0 team used this WhatsApp group for coordination purposes, as well as sharing and receiving data and information related to the issues in focus of the initiative (e.g. pictures and videos of biodiversity sighting and incidents or weather information). Face-to-face meetings were the place for discussions about the co-design process and different issues related to the CBM activities, meeting and discussing with representatives of different organizations and individuals, as well as receiving training on using the MMCO tools. Email was used for coordination purposes, mostly with organizations e.g. invitation to meetings. Only a few interviewees mentioned that they are using the MMCO App and website for sharing and receiving information.

Change in methods of communication and participation because of MMCO

Only a few interviewees believed that they were at the receiving end of communications during the participatory meetings and others mentioned that they had a chance to express their ideas during the participatory meetings. Several interviewees described MMCO as platform that is designed for data collection and sharing; a platform that has the potential to help them better understand their living environment, or issues related to their official mandate (e.g. conservation). Furthermore, MMCO was described as "instrumental to open conversation with some key stakeholders"[43]; a platform that enabled different stakeholders to meet and have open conversations, and also as a channel for establishing connections that can be used in the future. In addition, MMCO was also perceived as an educational tool. A school teacher mentioned that they use the example of MMCO in their lectures with students, talk about the types of data that is generated in the initiative and discuss how this is beneficiary for the local communities that live with the wildlife in the Mara. He mentioned that this will help the students to develop ideas and communicate the issues with their friends and family members. Comparing these findings with the baseline situation of this case, and the 'communication and decisions modes' identified by Fung (2006) and Wehn et al. (2015b) shows that MMCO has facilitated communication between different stakeholders, and allowed them to learn (listen as a spectator), participate in data collection and sharing, as well as express and develop preferences about biodiversity conservation and sustainable livelihood management.

Nevertheless, none of the interviewees in this case could identify an example of a mechanism by which MMCO has enabled them to (better) take part in decision making processes related to the issues in focus of the CBM. Interviewees often referred to the limited amount of data and information produced in MMCO as the reason for their answer.

A number of interviewees also believed that if the information produced in MMCO keeps accumulating, they will have a better understanding of their environment and, through that, they may be able to make better decisions within their organization on their community. An example of this situation was mentioned by a participant from one of the conservancies. He believed that the weather data and the information that is generated in

---

[43] KE-02-09

the MMCO App has the potential to provide their conservancy with a better picture of existing problems and helps them to develop solutions for those problems i.e. an increase in 'technical expertise' (Fung, 2006). He also mentioned that this will be very useful as it helps them to make better decisions within their conservancy, but also communicate the common issues with the neighboring conservancies and KWS.

5.3.3 Power dynamics in MMCO

This section presents the results of an analysis of the external and internal power dynamics in MMCO. This is directly linked to the third research question of this study i.e. who controls and influences MMCO and how?

Change in the social, institutional and political context of the case

Section 5.2.1 provided an overview of the social, institutional and political context of the Kenyan case study. Since this information was generated at the start of the establishment process of MMCO, the researcher was interested in capturing any changes in the social, institutional and political context, which is relevant for the issues in focus of the CBM.

Overall, the majority of MMCO members did not identify any major changes in the social, institutional and political context of this case study, since the beginning of the initiative activities in 2017.

The general election of 2017 at the national and county levels is perhaps the most important event that could have potentially affected the establishment process of MMCO. During this election, the Kenyans voted for electing their president, the parliament members, governors and the county assemblies. The presidential election in August 2017 was one of the most controversial elections in Kenya's history. The process of holding this election resulted in escalating violence and disputes across the country. In a historical verdict, the Supreme Court of Kenya nullified the results of the August elections and asked for a fresh election in October 2017. The nullification of the results was based on the flaws that were identified in the electoral process, including transmission, verification and transparency of the results (EU-EOM, 2018). Further detail about the process of these elections is beyond the scope of this research; however, the end results retained both the president Uhuru Kenyatta and the Narok County Governor Samuel Kuntai Ole Tunai in power and did not create a major change in the overall policies at the national or county level. Nevertheless, some ministers and department heads within the Narok County changed and this resulted in staff turnover within different departments of the county. This staff turnover and the fact that the new staff members were not a part of the co-design process of MMCO, or felt left out of this process, negatively affected the efforts for engaging the Narok County Government as one of the main stakeholders in this CBM. Moreover, after the elections, the management of the Maasai Mara National Reserve changed and a new Chief Park Administrator was appointed[44]. The Chief Park Administrator is responsible of overseeing the management of the park and revenue collection, as well as conducting research in the park. This change resulted in a much more restricted access to the park for research purposes; an issue that can negatively affect the amount of observations made within the national reserve, especially for those who want to use the MMCO tools for research purposes.

Establishment mechanism of MMCO

Similar to the Dutch case study, MMCO was also initiated as a project-driven CBM, in the context of the Ground Truth 2.0 project. The initial need for establishing this CBM was originated from the Ground Truth 2.0 partner organizations and based on their understanding of the local needs from previous projects and interactions that they had with local stakeholders in the Mara. Moreover, there were certain project requirements and initial framings (e.g. the fixed funding period and the initial focus on biodiversity monitoring) the influenced the objectives, establishment process, functioning and subsequently results of this initiatives. In addition, organizations and team members involved in establishing MMCO had certain expertise, preferences, resources and research interests that influenced the establishment process, functioning and results of MMCO. For example, designing an Android App in English was partly driven by preferences of Ground Truth 2.0 partners.

Although by definition, MMCO was co-created, based on a co-design approach (Conrad & Hilchey, 2011; Haklay, 2015; Shirk et al., 2012; Wehn et al., 2015a), the researcher wanted to understand the opinion of the interviewees about this process and therefore asked them for their views.

Interviewees from the MMCO members had diverse views about the establishment process of this initiative. Approximately, 35% of the MMCO members who were interviewed believed that MMCO was co-created and the stakeholders who were represented had the chance to influence its design and functionalities. Another 35% believed that this was a top-down process; a foreign idea or "yet another project"[45] that was brought to the stakeholders, but the process was mostly driven by ideas of the project partners or wishes of certain stakeholders. 20% of the interviewees from this group perceived the establishment process of MMCO as a bottom-up process that mainly included the ideas and wishes of the local citizens. The remaining 10% could not identify a specific establishment process for MMCO.

---

[44] KE-02-06
[45] KE-02-05

The disagreement on the establishment mechanism of this CBM was not limited to MMCO members and the three interviewees from the GT2.0 team had different views on its establishment process. Two interviewees believed that MMCO was co-created; however, the way they described the establishment process was fundamentally different. One of the interviewees believed that MMCO was established in a collaborative way by a group of interested stakeholders who had the chance to influence its design and functionalities. The other interviewee believed that GT2.0 team members "pre-empted the discussions a lot in the meetings" and influenced the functional design of MMCO by introducing some requirement on behalf of the stakeholder[46]. The third interviewee from this group believed that the establishment process of MMCO was bottom-up, it was informed by "the needs on the ground". This interviewee believed the government organizations were not very willing or committed to the cause during the establishment process[47].

All in all, the diverse views on the establishment process of MMCO shows that there is no common understanding about the establishment process of this initiative and may indicate that internal power dynamics in this case are seen differently by different individuals. Comparing these diverse views against the stakeholder categories that were interviewed did not yield any meaningful conclusion. Nevertheless, based on the researcher's observations, some of these diverse views may be explained by limited understanding of some of the interviewees about the concept of co-design. In addition, that fact that this CBM was project-driven and the GT2.0 team members had more control over the establishment process can explain some of the aforementioned views.

Change in access to and control over data because of MMCO

At the time of conducting this research, the MMCO platform provided access to data from 13 TAHMO weather stations, four of which were installed as a part of the Ground Truth 2.0 project. In addition, weather and water level data from a number of low cost weather stations and water level sensors, which were installed as a part of the MaMaSe project, was also integrated into the MMCO platform.

By the end of November 2019, 232 observations were submitted using the Mara Collect App. Figure 5.6 shows the temporal and thematic distribution of these observations from the first submission in March 2018. As the graph shows, not much data was produced using the App until September 2019. The peak of data collected that approximately represents 75% of the total observations relates to a coordinated data collection event from 24 to 26 September, during which a number of volunteers, mostly students from the Maasai Mara University and a number of high school students and teachers, were trained

---

[46] KE-02-02
[47] KE-02-01

126

on how to use the App. This event included a trip from Narok to the Maasai Mara National Reserve; a route which pretty much defines the spatial distribution of the observations (Figure 5.7).

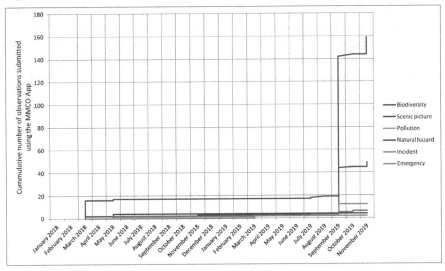

Figure 5.6 Temporal and thematic distribution of the data submitted using the Mara Collect App in 2018 & 2019
Source: (Author)

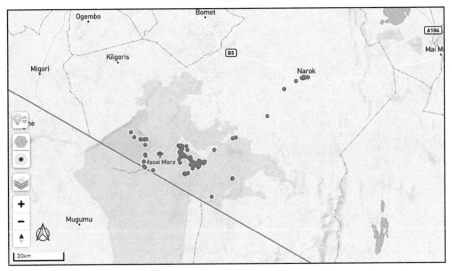

Figure 5.7 Spatial distribution of the data submitted using the Mara Collect App in 2018 & 2019
Source: https://ona.io/

127

In addition, mapathons were used as a participatory approach for increasing access to and control over data in this CBM. These are the coordinated mapping events, which took place during the lifetime of the Ground truth 2.0, with the aim of improving coverage of the maps in vulnerable places. These events were mainly coordinated by IHE Delft Institute for Water Education, but also at later stages by Maasai Mara University. Participants in these events were mainly students from different universities inside and outside the case study area, as well as GIS experts and other interested volunteers. Table 5.2 provides an overview of the mapathon events that happened as a part of MMCO.

Table 5.2 Overview of the mapathon events in MMCO

Date	Locations	Details
13/Feb/17	IHE Delft, the Netherlands Wageningen University, The Netherlands ITC, The Netherlands Upande, Kenya Other volunteers	200 participants
15/Nov/18	Maasai Mara University, Kenya Upande, Kenya MapKibera, Youth Mappers, IHE Delft, the Netherlands	2413 edits: 2061 buildings, 138 km roads *Number of participant not known*
28/Nov/18	IHE Delft, the Netherlands IHE Delft alumni in Uganda	150 participants
13/Mar/19	Mini mapathon at the community of practice workshop Open Water Network event in Arusha (Tanzania)	10 participants
28/Mar/19	Maasai Mara university, Kenya IHE Delft, the Netherlands	Online mapping for Cyclone Idai *Number of participant not known*
18/Nov/19	Maasai Mara University	30 participants

At the time of conducting this research, the Ground Truth 2.0 partners did not use the data/information produced in MMCO to a great extent and their use of data/information produced in this CBM was limited to research and education purposes, e.g. the use of data produced in Mapathons for teaching GIS. Similarly, most of the CBM members mentioned that they have not so far used the data/information produced in MMCO. The majority of MMCO members highlighted that the available data/information is very limited and therefore perceived little or no change in access to data/information. The only exception was the weather data that was used by some stakeholders, including lecturers and students at the Maasai Mara University who used it for education and research purposes, and the Meteorological Department of the Narok County that used this data for producing forecasts and alerts.

The Ground Truth 2.0 partner organizations believed that the control over data should stay with the local stakeholders and the data management policy of MMCO should regulate issues such as access to raw data as well as quality control or data sharing processes. They perceived the data policy as the main factor that defines which stakeholder(s) will have control over data.

The majority of MMCO members did not perceive a change in their control over data/information. The only exception was increased access to raw weather data from the weather stations that was only provided to some stakeholders, e.g. a few lecturers at the Maasai Mara University. Moreover, there were opposing views on issues related to control over data. An interviewee from the university believed that MMCO is an open system, and if needed, all members should be able to access data and process these. In contrast, an interviewee from KWS believed that control over data should be centralized within a relevant organization and this organization should be in charge of quality control and filtering of the data. This is linked to the discussions about "sensitive data' that resulted in producing a data sharing policy in this CBM. Interestingly none of the MMCO members referred to the data management policy as a determining factor for defining the level of control over data for different stakeholder. This, among other things portrays a low level of internalization of the data policy by the stakeholders.

Change in the authority and power of different actors because of MMCO

As part of this study, the researcher was interested in understanding any change in the levels of authority and power of different stakeholders as result of their participation in this CBM. In order to do so, interviewees from both the GT2.0 team and the MMCO members were asked; to what extent, if any, they think their influence in decision making processes regarding the topics of biodiversity conservation and/or livelihood management in the Mara region has changed because of their participation in MMCO.

The GT2.0 team members provided examples of change in authority and power of their organization, which based on Fung (2006) includes 'personal benefit', 'communicative influence' and 'advice and consult' modes. For example, the representative of one of the partner organizations mentioned that they are now "on everybody's radar"[48], and as a result, their company has been invited to a number of meetings for discussing the Spatial Plan and the Integrated Development Plan of the Narok County. Another interviewee from a different partner organization mentioned that they have been successful in communicating the value of data for decision making processes to a number of local and national organizations and they believe that if those stakeholders want advice or consultation with this regard, they will consult them[49].

Compared to the Ground Truth 2.0 partner organizations, this change was much less evident for the MMCO members. The majority of interviewees from this group did not perceive any change in their level of authority and power as result of their participation in this initiative. Those interviewees who already had a mandate for making decisions (e.g. interviewees from the Narok County, KWS, or KFS), believed that regardless of

---

[48] KE-02-01
[49] KE-02-03

MMCO, they have a role in, and a mandate for, making decisions regarding biodiversity conservation and/or livelihood management. In addition, approximately half of interviewees who did not have any expectation to influence policy and decision making processes believed that MMCO has not helped with changing their influence. Nevertheless, the other half perceived a change in authority and power for influencing decisions that resulted from participation in MMCO. Some of these interviewees mentioned that through participation in the meetings and through using the MMCO tools, their knowledge has increased and they are now aware of the fact that they can contribute to observations and use this information for their own purposes. Some interviewees also mentioned that due to this increased knowledge and awareness, they are now better placed to influence public opinion and educate other community members. Comparing these finding with the baseline situation that was presented in section 5.2.1, demonstrates that MMCO has not created new ways for exerting influence on decision making processes, rather it has reinforced the already existing methods by increasing participants' knowledge and awareness.

Revenue streams of MMCO

Based on the report about the sustainable Business models for the Ground Truth 2.0 products/services, there are three possible options for sustaining MMCO (Kersbergen et al., 2019).

Option 1: Upande, as the local partner of the Ground Truth 2.0 project and the main developer of MMCO tools, receives a contract for hosting the MMCO platform and to maintain it. Nevertheless, it is not clarified who will pay for the aforementioned contract.

Option 2: Maasai Mara University will host the MMCO platform, with close support from the Narok County Government and MMWCA. This means that the three aforementioned organizations are responsible for operation and maintenance costs of MMCO. In this scenario, a period of one month is envisioned for handing over the tasks to other parties who will be involved in the future operation and maintenance of MMCO.

Option 3: The African Conservation Centre will host the platform, when and if a decision is made that the MMCO services will be supported at national level.

The aforementioned report identifies the first option as the most probable modality to sustain the MMCO in the future and mentions that in this scenario, Upande will be contracted by the Narok county or the African Conservation Centre to continue providing their services.

In terms of revenue streams, all three envisioned options indicate a 'sponsorship' model (Gharesifard et al., 2017, 2019b), which requires the commitment of organizations such as the Narok County Government, African Conservation Centre and/or Maasai Mara University to provide the required financial, technical and organizational support to

131

sustain MMCO. However, by the end of November 2019, no official commitment was made by the mentioned organizations.

The results of the interviews also demonstrate that there is no common understanding about the future sustainability of MMCO among the stakeholders. When asked about the envisioned revenue streams for sustaining MMCO, almost all interviewees from both groups mentioned that this is not (yet) very clear and it is an ongoing discussion by the MMCO members. Nevertheless, several interviewees suggested possible funding sources for MMCO. Some of these interviewees suggested that the funding should come from one or more organization involved in establishment of MMCO. Examples mentioned included the Narok County Government, KWS and Maasai Mara University. External funding sources such as donors (e.g. USAID or UNDP) or national level organizations such as the Ministry of Education were also suggested possibilities. Not surprisingly, the majority of the interviewees suggested that the revenue streams should come from other organizations and barely volunteered their own organization. The only exception was Maasai Mara University that was willing to provide a venue for future meetings and provide technical support for future mapathon events. Moreover, a few interviewees were skeptical about the continuity of MMCO and believed that after the funding from the project finishes, the initiative activities may stop because different stakeholders still cannot clearly see the full potential or added value of MMCO.

5.3.4 Technological choices for MMCO

The content of this section, aims to answer the fourth research question of this study for MMCO (i.e. How effective and appropriate are the choices and delivery of the selected technologies of MMCO?). This section includes an overview of the technological components used in MMCO, assessment of accessibility of these technologies and how these relate to existing infrastructure, as well as discussions about included and excluded groups as result of technological choices.

Technologies used in MMCO

Similar to the Dutch case, the technical design of MMCO was also informed by its functional design processes and a users' story map that is presented in Figure 5.8.

The story map analysis in MMCO resulted in identifying two main categories of functionalities, which then guided the technical design of this CBM; (1) support collection and sharing information about the topics of biodiversity conservation and sustainable management of livelihoods in the Mara, and (2) provide channels for communication among different stakeholders that can be used for consultation and planning purposes (Omoto et al., 2018).

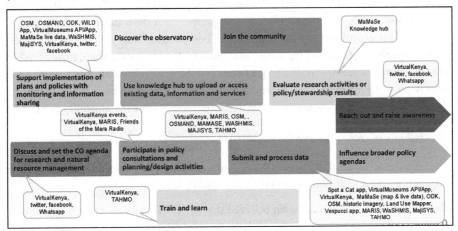

Figure 5.8 Story map of MMCO
Source: Omoto et al. (2018)

The technological components of MMCO include newly developed or purchased components, as well as added components from existing technologies.

The newly developed or purchased components include Mara Collect and the MMCO Apps, the Maasai Mara Citizen Observatory website that uses a virtual (Amazon) server, four TAHMO weather stations and four screens that were installed as a part of the Ground Truth 2.0 project. Technical design choices in developing the MMCO Apps were mainly driven by preferences of the Ground Truth 2.0 project partners. More specifically, Ground Truth 2.0 partners in charge of developing the tools had a strong influence in making technical design choices. Some of these choices matched the local context very well, while others did not. The Mara Collect App uses ODK; an open-source software that allows for collection of data without internet connectivity; a design choice that fits the technological context in which MMCO was established and also has lower maintenance costs than non-open source options. Nevertheless, as a for-profit company, Upande was interested in providing convenient and cost-effective technical solutions that can be rehashed and repurposed in future. This led to design choices such as use of English language and design of the Apps only for Android devices. Annex 7 presents screen shots of MMCO's web-platform and Apps.

Added technological components include the data feeds coming from nine other TAHMO stations, and MARIS that includes the data from a number of low cost weather stations and water level sensors from the MaMaSe project. The water level and weather stations form the MaMaSe project operate with a SIM card technology and are maintained by Upande.

The next point about the technologies used in MMCO relates to the issue of ownership of technologies. Some of these technologies are owned and operated by specific project partners and access to these technologies is pretty much determined by those project partners. This especially applies to the added components from existing technologies, such as weather stations and the water level sensors that are operated and maintained by the technological partners of the project (i.e. TAHMO and Upande). It is yet unknown who is going to maintain these technological components and who is going to pay for these services in the future; uncertainties that threaten the future sustainability of MMCO.

Accessibility of technologies used by MMCO

The interviewees from the Ground Truth 2.0 team and the MMCO members had opposing views about the accessibility of technologies used in MMCO.

Overall the Ground Truth 2.0 team believed that the CBM tools are fairly accessible for a large part of the population and especially those who know English and have access to smart phone and internet. They also mentioned that the cost of purchasing a smart phone is dropping (e.g. at the moment you can have one for around 40 dollars) and the design of the MMCO App allows for offline data capturing, which helps a lot with its use in places with weak or no network coverage. However, they believed that a simple instruction session is required in the beginning to familiarize the users with the practical steps in using the App.

The majority of the MMCO members, in contrast, mentioned that the App and the website are not accessible enough for an average local community member, especially for local pastoralists or farmers. Limited access to smart phones and internet or the willingness to use these, use of English language, as well as low average level of literacy and digital skills were among the factors that limited the accessibility of the tools. It was mentioned that people need training before they can start using the App and therefore currently mostly the people who participated in the meetings or training sessions are able to use the App. There is no manual or instruction available for those who may want to learn using the App on their own. Furthermore, an interviewee mentioned that "local people in the Mara don't use smart phones because smart phones go out of battery soon. Community member and rangers who want to spend a few days in remote areas prefer to have a normal phone that holds battery for a long time"[50]. Some interviewees also highlighted the need for simplifying the App to make it more accessible for the less literate community members or those with a lower level of digital skills. Use of pictures and illustrations instead of text was one of the suggestions for making the App more accessible for community members. Moreover, it was mentioned that people do not often use websites (an issue that was also identified in the baseline analysis of this case); therefore the information shared on the website may not be viewed as much as the tool developers may have hopped. The cost of data bundles was identified as another barrier for accessibility of the tools. Interviewees highlighted that people need to pay for data bundles, in order to be able to use the tools and sometimes even the rangers are not compensated for the use of their data bundle to use such an App.

Included and excluded groups resulting from technological choices

Developing an App, based on ODK technology, in a geographic location with poor internet connectivity such as this case really fits the context and allows for a more widespread use of the technology. However, developing an App-based CBM in a region where a large percentage of the population do not own a smart phone or do not have the technical skills to use it may result in exclusion of these groups from using the App and benefiting from it. In addition, the MMCO Apps are only available for Android devices, an issue that excludes end-users who use devices with other operating systems such as iOS.

In order to submit an observation, users need to be able to read and write in English, which is not the case for many local stakeholders who only speak a local language like Swahili. Moreover, the design of the Mara collect App is pretty-much text-based and users need to navigate their way through a series of lists and options that all appear in a text format. This means that the users need to have a certain level of literacy, a skill which is lacking in a large part of the population, especially the older generation. With this regard a interviewees mentioned "if we have the possibility to have the information in a picture form, more community members can access and use the information"[51].

5.3.5 Results of MMCO

Interviewees in this case were asked to identify the realized and expected outputs (direct products) of MMCO. They also expressed their perception about the realized outcomes (actual short-term or incidental changes) that have happened because of MMCO, and the outcomes that can be expected to happen in the near future. This is directly linked to the

---

[50] KE-02-16
[51] KE-02-04

fifth research question of this study (i.e. what are the expected and realized outputs, outcomes and impacts of the CBM initiative?). Impacts of a CBM take a long time to take place and their study was out of the scope of this research, nevertheless, interviewees were asked to express their expectation of long-term changes that may happen in the future a result of MMCO. Similar to the Dutch case study, interviewees in this case were also asked to think of both positive and negative outputs, outcomes and impacts. However, uncertain future outcomes and impacts, and 'social desirability bias' (Fisher, 1993) because of interviewees affiliation with MMCO, may have influenced the responses. The results reported in this section summarize the expected and realized about the outputs, outcomes and impacts of MMCO.

Outputs of MMCO

The technological components of MMCO, including the two Apps, the website, as well as the weather stations and screens that were installed, were the most tangible realized outputs (direct products) of MMCO for the local stakeholders. In addition to these tools, the interviewees from the GT 2.0 team also elicited the added technological components from existing tools and services (see section 5.3.4), and also the creation of the MMCO WhatsApp group (with some 60 members), as the main technology-related outputs of MMCO. The data including the photos and observations submitted using MMCO App, weather data, and the maps produced during the mapathon events were also identified by the CBM members as a direct product of MMCO, however, interviewees often described this data as 'dispersed' or 'occasional'.

Another identified output of MMCO was the educational material related to mapathons and weather stations that can be used for teaching and training purposes. The Maasai Mara University and the schools found immediate benefits for developing their education and research curriculum, using mapathons, the data from the weather stations and the functionalities of the App.

Although the data sharing policy produced in MMCO can be considered as an output of this initiative, none of the interviewees identified this as an output for MMCO.

Increased engagement of stakeholders and a better uptake of the MMCO tools, which results in improved datasets on biodiversity, meteorology, livestock, etc., was mentioned quite frequently as what interviewees wished to see as the outputs of MMCO. Interviewees highlighted that they anticipated more efforts on capacity building, awareness raising, training and outreach, and they expected to see more or better engagement of some stakeholder, especially government agencies (e.g. NEMA, Water Resources Authority and Drought Monitoring Authority), NGOs, as well as local communities such as pastoralists and farmers. Some interviewees from the Narok County Government, KWS and conservancies believed that MMCO has the potential to provide a lot of information that are relevant for a large number of organizations, but currently

136

this information is difficult to use. They described the data and information produced by MMCO as generic and unstructured and mentioned that they expected to see this data and information in a more processed and organized way. One of these interviewees used the term 'clean data' to describe this, and defined 'clean data' as data that has a theme, is vetted, categorized and is usable for different purposes[52]. With regards to usability of data for different organizations, a few interviewees mentioned that perhaps in addition to the discussions in the co-design meetings, bilateral need assessments, targeting different organizations, should have been conducted by the Ground Truth 2.0 project to make the output data/information useful for their purposes.

Outcomes of MMCO

Interviewees from both groups also identified multiple community-related outcomes for MMCO. These outcomes mostly related to changes for individuals or organizations who participated in the MMCO workshops and training sessions. Increased awareness about the concept of community-based monitoring, establishing a community of stakeholders with a shared vision and mission, creating knowledge and awareness about the fact that data gaps exist, and creating an understanding about how this can be tackled using a participatory approach, with inclusion of all stakeholders, were among these identified outcomes. Nevertheless, some interviewees believed that this knowledge and awareness raising has happened at a small scale, and mainly within those who participated in the MMCO meetings and training, and there is still a need for engagement and outreach to a larger number of community members.

Quite a few interviewees envisioned future outcomes for MMCO, but almost all of these interviewees mentioned that realization of these outcomes depend on the continued involvement of already engaged stakeholders and increased uptake of the CBM activities by more people. One of these envisioned outcomes was learning at the individual and society levels that can result in change in attitude towards important issues such as conservation. MMCO was also envisioned as a channel through which community members, government organizations and researchers can communicate and exchange data and information. This process was seen as a mechanism that can generate scientific outcomes and a better understanding of the environment, which in turn can be used by different organizations (e.g. by the Narok County, KWS or conservancies) for better-informed decision making. In addition, MMCO was seen by some interviewees as a platform that can help community-members' voice be heard by decision makers, e.g. on the issue of human-wildlife conflict. Moreover, it was also mentioned that MMCO can

---

[52] KE-02-16

help improve the economy at the individual and societal level, e.g. through providing information on the market prices for livestock, grazing areas and water resources.

Expected impacts of MMCO

Envisioned governance-related impacts of MMCO included evidence-based environmental decision and policy making and conservation actions, engaging community member and giving them a voice in decision making processes, planning for uncertainties such as climate change, as well as facilitating more transparent and accountable environmental governance.

Economic impacts were identified at both individual and societal levels. For example individual pastoralists can benefit from up to date livestock market prices and both the pastoralist community and the farmers can utilize the generated information to changing their practices e.g. by finding better pasture for their livestock or by accounting for climate change mitigation measures in future agricultural planning.

Some of the interviewees also linked these to positive environmental impacts such as avoiding overgrazing, reforestation, reducing human-wildlife conflict and conservation of biodiversity and other natural resources.

5.4 DISCUSSION

The results of the baseline analysis of the Kenyan case study (section 5.2 above) and a systematic evaluation of the establishment process and results of MMCO (section 5.3 above) enabled answering research questions 1 to 5 for this CBM. This section is therefore dedicated to presenting the answers to these five questions and discussing how these findings relate to previous research about the establishment and functioning of CBMs. Moreover, answers to the five critical questions posed by the CPI Framework helped clarifying the meaning of community in MMCO; a discussion that is presented and elaborated in this section.

RQ1. What are the overarching objectives and actor-specific goals of Maasai Mara Citizen Observatory and to what extent does the design of this initiative help achieve those goals/objectives?

According to the vision and mission of MMCO, this CBM aims at promoting environmental monitoring and environmental stewardship by facilitating data collection and sharing, and by supporting collaboration of all stakeholders for creating a balance between sustainable livelihoods and biodiversity management in the Mara ecosystem. This vision and mission aligns well with constitutional and legislative recognition of

138

public participation as a national value and governance principle, as well as the right of access to data for the general public in Kenya (Government of Kenya, 1999, 2010, 2013).

Using a co-design approach, the members of this CBM jointly defined four objectives for this initiative (see Table 5.1). These objectives were defined quite broadly and did not change throughout the establishment process of MMCO. In essence, with these four objectives, CBM members aimed at developing a socio-technical platform that contributes to increased access to a wide range of data and information, and support better decision and policy making, which in turn would lead to a better balance between biodiversity conservation and livelihood management. By November 2019, the objectives of MMCO were only partly achieved. Several reasons were identified for the partial achievement of the objectives, which included low level of engagement and actual use of tools, lack of sense of local ownership among local stakeholders, inadequacy of communication and outreach efforts by the CBM members to disseminate the idea to others, and social, institutional and technological constraints.

Limited access to mobile phones and internet (Intelecon, 2016; ITU, 2017b), as well as low level of interest in digital technologies among certain groups within the community (e.g. elderly) contributed to low level of engagement and use of the MMCO tools. Moreover, social constrains such as the widely practiced norm of receiving DSAs for participation in meetings and low level of literacy among general public (Ngugi et al., 2013) hindered both offline and online participation in MMCO.

Another contributing factor relates to sensitivity of some data and information produced in this CBM. Data sensitivity is a common challenge for many CBM initiatives (Newman et al., 2011), especially those focusing on information about threatened and endangered species (Crall et al., 2010). While the tools for collecting data were in place, the absence of an agreed-upon data sharing policy hindered rolling out MMCO activities at a larger scale and this negatively affected engagement momentum in this CBM.

As indicated by Bonney et al. (2009a) and Shirk et al. (2012), following a co-design methodology for defining the objectives and functionalities of a CBM allows for inclusion of diverse stakeholder-identified needs. The four co-designed objectives of MMCO aligned well with the identified actor-specific goals in this CBM. Nevertheless, as argued by McElfish et al. (2016) a key challenge for CBMs is to determine how to interact with government officials in an efficient way and how to produce data and information that is useful for improving public decisions. In case of MMCO, engaging 'the right people' in the establishment process proved to be difficult. Some organizations sent junior staff to the meetings and they did not communicate the discussions to the higher levels of decision making in their organizations. Moreover, staff turnover in medium level management of the Narok County Government negatively affected the efforts for engaging them. In addition, because of the broad focus of this initiative, the data and information generated through it were interpreted as 'generic'. Therefore, some government organizations did not have an interest in using the generated data and information and considered these as un-structured and good to have, but not very useful for their organizations.

The low level of participation and actual use of tools and lack of sense of ownership among local stakeholders may be correlated with the fact that MMCO is a project-driven CBM. Although the issues in focuses of this CBM are very well-known and contested local issues, the initial need for tackling those issues via establishing a CBM was first introduced by the Ground Truth 2.0 project and in that sense was not demand-driven. Another contributing factor to the lack of sense of local ownership may related to the findings of Bartels et al. (2017) that depicts Maasai as an over-researched community. It has been argued that the large number of past research efforts in the Mara has not resulted in meaningful changes in livelihoods of local stakeholders (Bartels et al., 2017). Therefore, some local stakeholders may have perceived MMCO as yet another research project with no tangible benefits.

RQ2. Who participates in Maasai Mara Citizen Observatory and how?

MMCO allows for both offline and online participation of interested stakeholders. Offline participation is mainly via face-to-face meetings and outreach events, and online participation mainly happens through using the two APPs and the web-platform of this initiative.

Participation in MMCO can be divided to two phases. The first phase was the co-design phase in which this CBM was established and its functionalities were defined. Participants in the co-design processes included representatives of local and national level organizations (e.g. the Narok County Government, KWS, KFS, etc.), universities (Egerton University and Maasai Mara University), schools, NGOs (e.g. the African Conservation Centre and Friends of Maasai Mara and MMWCA), as well as individual community members and representatives of organized community groups. The second phase is the phase in which members of MMCO can utilize the co-designed functionalities for participation in this initiative. The co-design process of MMCO has resulted in creating a number of bi-directional communication and information sharing possibilities for the stakeholders. Online interactive dialogue and information sharing happens mainly using a dedicated WhatsApp group and less frequently via the CBM Apps. Forming a WhatsApp group for communication with and among the stakeholders was informed by the baseline assessment of existing patterns of communication and proved to be very effective for both coordination and information sharing purposes. Face-to-face interactions during the co-design and planning meetings provided opportunities for interactive dialogue among the CBM members. This created a change in communication modes (Fung, 2006) by facilitating awareness raising, communication, and data and information sharing between different stakeholders at a small scale.

The uptake of the tools in MMCO was mainly limited to the use of weather stations by universities, schools, and the Meteorological Department of the Narok County, as well as the use of mapathons for educational purposes. Data sharing by citizens using the MMCO Apps is very limited. Moreover, only a few users (not more than 24) regularly visited the web-platform of MMCO. Comparing these figures with the potential pool of end-users (more than 850,000 inhabitants in the area), and even the number of people trained for using the MMCO Apps (more than 70) shows that only a small percentage of potential end-users actually use the MMCO Apps and web-platform. Some of the reasons for this limited participation in MMCO are detailed hereafter.

The geographically dispersed population in the Mara and the large number of inhabitants in the case study area resulted in limited representation of local community members in the co-design process; an issue that negatively affected the broader uptake of MMCO tools. The importance of having central coordinators for engaging volunteers, facilitating CBM activities and information flow has been highlighted in previous research (Pollock & Whitelaw, 2005; Richter et al., 2018). Absence of central coordinators, key local influencers and community leaders such as representatives of the church (e.g. spiritual leaders) and local chiefs in the co-design process made it difficult to recruit more local community members. Given widespread geographic coverage and the dispersed population, recruiting local coordinators for facilitating the activities and disseminating the results could have contributed to a better achievement of MMCO objectives.

Previous research has identified availability of resources such as equipment, financial means and technical skills as influential factors on participation of stakeholders in CBM initiatives (English et al., 2017; Gharesifard & Wehn, 2016a; Gharesifard et al., 2017; Wehn & Almomani, 2019). Limited access to smart phones and internet, cost of data bundles, transport costs, time commitments required for online and offline participation and the need for training for using the MMCO Apps were identified as hindering factors for participation in MMCO.

Using complex and detailed data collection protocols may negatively affect participation and rate of data collection (Birkin & Goulson, 2015; Roy et al., 2012). The text-based design of the Mara collect App requires the users to navigate their way through a long list of options; this has negatively influenced the usability of the App. Moreover, design of the Apps in English language has dictated a certain level of literacy and language skills for its use and contributed to its limited uptake.

In addition, during the lifetime of the Ground Truth 2.0 project, technical, organizational and financial support required for the establishment and functioning of this CBM was provided by the project partners and through project funds. Nevertheless, from 2020, MMCO needs to sustain itself, while its future revenue streams are (yet) largely unknown.

141

RQ3. Who controls and influences Maasai Mara Citizen Observatory and how?

Due to the fact that MMCO is a project-driven CBM, the Ground Truth 2.0 partner organizations and the team members involved in establishing this CBM had a certain level of control over, and influence in, its establishment process. Accordingly, availability of resources and expertise within these organizations, project needs, budgetary requirements and promises made to funders were among influential factors that shaped this CBM.

MMCO was established using a co-design methodology. On the one hand, it has been argued that co-created CBMs should strive for balancing power relations between different actors (Eleta et al., 2019), and on the other hand CBMs that focus on environmental governance issues interact with inherently political processes (Cleaver, 1999) and include existing power dynamics between different stakeholders (Newman et al., 2012; Wehn et al., 2015b). Similarly, it has been argued that "making citizens more central within the science-policy process is inevitably constrained by pre-existing uneven power relationships between politicians and citizens, scientists and citizens, and scientists and politicians" (Kythreotis et al., 2019, p. 6). MMCO interacted with the existing power relationships between national and local level authorities, local community members, other stakeholders (e.g. schools, universities), and even the GT2.0 team members. Existing power relationships between the aforementioned actors affected decisions such as possibilities for data collection and sharing and the design of the tools in MMCO, which in turn affected the uptake and shaped the results of this CBM.

The establishment process of MMCO followed the same co-design methodology as Grip op Water Altena, however, power dynamics during the establishment process was perceived differently by different members of this initiative. Some CBM members described the establishment process of MMCO as a co-designed process, some as a top-down process mostly driven by GT2.0 member or wishes of certain stakeholders, and others as a bottom-up CBM that mainly reflects the ideas and wishes of the local citizens. Comparing these diverse views against the stakeholder categories did not yield any meaningful conclusion. Disagreement on the establishment mechanism of MMCO shows that there is no common understanding about the establishment process of this CBM and indicates that internal power dynamics in this case are seen differently by different individuals. Limited understanding of some of the interviewees about the concept of co-design and control of the GT2.0 team members on establishment process can help explain these differences in perceptions.

There was an agreement about limited change in access to data because of MMCO among different stakeholder groups. The most frequently mentioned increase in access was related to the data from weather stations. Sensitivity of some data and information produced in this CBM called for having a data sharing policy. The need for this data sharing policy inevitably maintains certain pre-existing power relationships between authorities and citizens (Kythreotis et al., 2019) by keeping the access to and control over a part of data and information with government organizations such as the Narok County Government, KWS and KFS.

MMCO did not contribute to a change in the level of authority and power of different stakeholders; however, it reinforced already existing possibilities for local community member to exert communicative influence (Fung, 2006) on decision making processes. The result of the baseline analysis of this case is in agreement with findings of a recent research by Sauti za Wananchi (Voices of Citizens) that identified attending public meetings and expressing ideas and concerns as the predominant method for citizen participation in (environmental) governance in Kenya (Sauti za Wananchi, 2018)[53]. MMCO facilitated online and offline interactions among stakeholders, and increased the knowledge and awareness of its members regarding the topics of biodiversity conservation and livelihood management.

During the lifetime of Ground Truth 2.0 project, the financial means, as well as the organizational support for establishing MMCO was provided by the project (Wehn et al., 2020). The envisioned revenue streams for future sustainability of this CBM are not (yet) clear. Government sponsorship model is the most probable option, but there is no agreement or official commitment from CBM member organizations to provide this financial support. Stakeholder(s) that provide the financial support for continued activities of MMCO are expected to have a stronger say in its future functioning.

RQ4. How effective and appropriate are the choices and delivery of the selected technologies in Maasai Mara Citizen Observatory?

The main technological components of MMCO are two Apps, a number of physical sensors and a web-platform that enable collection and sharing of data and information about the topics of biodiversity conservation and sustainable management of livelihoods in the Mara.

Gouveia and Fonseca (2008) and Newman et al. (2012) warned about inadvertently widening the 'digital divide' gap between those who own or adopt the technologies developments in a CBM and those who avoid it or lack the required skills to access those technologies. Accessibility of technological components of MMCO is perceived differently by the GT2.0 team members and CBM members. GT2.0 team members believe the aforementioned technological components are fairly accessible for a large part of the population. Decreased cost of purchasing a smart phone during the past few years and the design of MMCO App that allows for offline data capturing were among the main reasons for this claim. On the contrary, CBM members believe that technologies are not

---

[53] The meetings that this research refers to include any public meeting (except for religious meeting) and the organizers of these meeting could be governmental or civil society organizations or any other actor.

accessible enough for an average local community member. The main elicited reasons for low accessibility of technological components were limited access to smart phones and internet (Intelecon, 2016; ITU, 2017b), the costs of data bundles, low average level of literacy and digital skills among the local community members (Ngugi et al., 2013), the text-based design of the App, use of English language, and the need for an initial training for using the Apps.

As a result, MMCO may exclude quite a few groups from using its technological components and functionalities. It has been argued that exclusion from participatory processes happens due to heterogeneity of communities (Cleaver, 1999) and often excludes the weaker strata in society (Flyvbjerg, 1998; McGuirk, 2001). In the case of MMCO, non-English speaking local community members, illiterate people, local community members without a smart phone or the technical skills required to use one (including a large part of community members from older generation), and people who cannot use a Smartphone while in the field for a prolonged time without being able to charge the phone are among the groups most likely excluded.

RQ5. What are the expected and realized outputs and interim outcomes of Maasai Mara Citizen Observatory?

Technological components including the two Apps, the website, as well as the weather stations and screens were perceived as the most tangible realized outputs of MMCO. Moreover, added components from existing tools and services, the WhatsApp group of MMCO, produced data and information in this CBM and the educational material related to mapathons and weather stations were also among the mentioned realized outputs. However, engagement of more stakeholders, a better uptake of the MMCO tools, and generation of less 'generic' and more structured and targeted data outputs was expected by both the GT2.0 team and the CBM members.

Previous research has shown that publishing sensitive biodiversity data undermine conservation efforts (Pimm et al., 2015; Tulloch et al., 2018). Sensitive data produced in MMCO raise concerns about the potential misuse of this data and may be a negative output of this CBM. For example, poachers can potentially abuse the data produced in MMCO for locating endangered species. The data sharing policy of this CBM is the envisaged mechanism for preventing such misuse. Although the data sharing policy is necessary for preventing such negative outputs, it cannot not guarantee that the sensitive data and information produced in MMCO is not leaked (e.g. by those who have access to sensitive data). This raises an ethical issue about the data produced in MMCO.

Knowledge and awareness raising (at a small scale) about the concept of community-based monitoring and also about the existing data gaps related to biodiversity conservation and sustainable management of livelihoods was among the realized outcomes of MMCO. Moreover, this CBM succeeded in creating a small community of stakeholders that aim for creating a balance between biodiversity conservation and sustainable management of livelihoods in the Mara through facilitating interactions and dialogue among stakeholders, by creating possibilities for contribution to environmental observations, and via promoting environmental stewardship. Expected future outcomes largely depend on continued involvement of already engaged stakeholders and increased uptake of the CBM activities by more people. Nevertheless, at the time of conducting this research, absence of the a finally agreed sustainability plan beyond the lifetime of Ground Truth 2.0, small number of engaged and active CBM members, and limited uptake of its tools does suggest that this CBM may discontinue in the near future.

The impacts of a CBM initiative take a long time to take place (i.e. based on Phillips et al. (2014), a period of 5-10 years) and MMCO is not an exception. Therefore, at the time of conducting this research, the impacts of this initiative was largely unknown however, if MMCO continues and successfully engages a larger number of community members and other stakeholders in the Mara, it has the potential to bring about positive impacts. Identified expected future impacts by the interviewees in this case were mainly governance-related, environmental and economic.

Governance-related impacts of MMCO include contribution to evidence-based environmental decision and policy making and conservation actions in the Mara, engaging community member and giving them a voice in decision making processes regarding biodiversity conservation and sustainable management of livelihoods, planning for uncertainties such as climate change, and facilitating more transparent and accountable environmental governance in this region.

Possible environmental impacts included avoiding overgrazing, reforestation, reducing human-wildlife conflict and conservation of biodiversity and other natural resources. Through these positive environmental changes, this CBM also has the potential to contribute to improvement of the economic situation for both individuals (e.g. local pastoralists and farmers), and different sectors of the local economy, e.g. agriculture or tourism.

The aforementioned anticipated future impacts of MMCO are broad, ambitious and arguably exceed the expectations from CBM platforms. The findings of Warner (2006) and Leeuwis et al. (2018) question the ability of CBM platforms to help address complex environmental challenges. Warner (2006), states that multi-stakeholder platforms barely have a significant mandate in governance processes and decision making power is rarely shared or devolved via such platforms. Leeuwis et al. (2018) argue that dependencies with different governance levels, communities, and other path dependencies (e.g. political, biophysical and historical) does not allow CBM platforms to meaningfully contribute to management of common resources. Nevertheless, CBM platforms are believed to be well-positioned for fostering co-creation of data and information and knowledge, awareness raising among stakeholders and facilitating communication and connectivity among stakeholders (Leeuwis et al., 2018); ambitions that are better aligned with the realized outputs and outcomes of MMCO.

What constitutes 'community' in MMCO?

The community in MMCO consists of a small number of members who have an interest or stake in biodiversity conservation or sustainable livelihood management in the Mara. The shared aim of this community is to contribute to creating a balance between biodiversity conservation and sustainable livelihood management in the Mara. The issue in focus of this CBM is very broad, complex and sensitive. This community and its vision, mission and objectives were co-created using the Ground Truth 2.0 co-design methodology. The issue in focus of this CBM was initially informed by existing insights of the GT2.0 partners into the local problems and the needs of local stakeholders. Nevertheless, the idea of, and the need for, establishing a CBM for tackling these problems did not originate from the local stakeholders. Rather it was project-driven and inevitably was influenced by certain limitations and pre-framings such as a fixed timeframe for establishment, expertise of partner organizations and GT2.0 team members, as well as a pre-defined funding and scope. The number and composition of MMCO members changed throughout time, but never reached a critical mass. Core members of community in MMCO include representatives of the Narok County Government, KWS, KFS, universities, schools, NGOs, and a small number of local community members who participated in its co-creation process. Community members in MMCO interact both offline, during face-to-face meetings, and online using the tools developed in this CBM. MMCO members could not agree on a way forward that guarantees the financial and operational sustainability of MMCO. Therefore, it is highly likely that MMCO activities stop in the future, and as a result, the community in MMCO shrinks in size or completely disappears.

5.5 CONCLUSIONS

This chapter presents the findings of the baseline analysis of the Kenyan case study, as well as the establishment process and results of MMCO. This is in line with the second and third objective of this research that is testing the empirical applicability of the conceptual framework and evaluating the evolving processes, outputs and outcomes of CBMs over time. The following conclusions sum up the most important insights generated from studying this case. It is important to mention that similar to the section 4.4, this section does not aim at generalizing the conclusions from studying this particular case. Rather, it is meant to present critical insights that can inform the establishment processes of other CBMs or studies that aim at understanding such processes.

Defining broad objectives and focusing on monitoring multiple environmental attributes increases the chance of being appealing for a large number of stakeholders, but at the same time makes it difficult to produce data and information that are specific enough and match different stakeholders' needs. This is especially important for CBMs that aim at improving public decision making processes. Moreover, achievement of the objectives of a CBM may be influenced by the social and technological contextual factors, existing ways of working within organizations, as well as data sensitivity issues.

Establishing CBMs using a co-design methodology often involves participation of stakeholders in multiple face-to-face or online meetings. Organizing co-design meetings requires funds and may involve a lot of travelling time for participants, which is especially a challenge in areas with geographically dispersed population. Engaging local influencers, community leaders and local coordinators may help reduce engagement efforts and help with the broader uptake of such initiatives. Limited access to technological components (e.g. smart phones and internet), complex data collection protocols and absence of resources such as time and financial means for participation are among the factors that undermine the efforts for recruiting members in a CBM. Therefore, there is a need for an early stage study of contextual realities and design factors that may affect participation in a CBM.

Although co-designed CBMs strive for creating equal opportunities for different stakeholders to influence the establishment process, interaction of these initiatives with existing and often un-even power relationships between different actors is inevitable. Interventions such as CBM projects should therefore be power-sensitive in their process of establishment. This means facilitator should actively consider and reflect on questions such as which groups are overrepresented, excluded, heard or ignored? Whose interests are more/less represented in the design of functionalities? Or which existing actor relationships are changed, kept or reinforced?

Technological developments for a CBM can both enable and constrain participation of groups within society. Establishing CBMs in developing countries and regions with less technological advancements is particularly more challenging and requires careful considerations for inclusion of vulnerable and less tech-savvy community members. Compatibility of technological choices with social, institutional and technological contexts reduces the chance of excluding major group within society. Nevertheless, heterogeneity of society should be acknowledged and choices need to be made about the extent to which CBMs can enable participation of different groups within society. Moreover, in the case of project-driven CBMs, there is a high chance that existing interests and expertise of initiating organizations or additional control of project facilitators influence the technological developments.

Data, information and knowledge exchange, awareness raising and facilitating communication among stakeholders are often among the results of CBMs that are immediate and easier to study. On the contrary, other outcomes and impacts of a CBM such as contribution to change in actor relationships or solving complex environmental challenges take time to materialize and become tangible or measurable, and are difficult to study. The timeline of this research did not allow for studying the medium and long-term changes resulting from MMCO. Nevertheless, the extent to which CBM initiatives can contribute to solving complex environmental challenges or balancing existing and un-even power relationships is often limited and therefore should not be overestimated.

6

CROSS CASE ANALYSIS[54]

Chapters 4 and 5 of this dissertation presented the findings and conclusions of the two cases studies of this research. Objective 4 of this research was to provide recommendations for CBMs based on a detailed analysis of the characteristics of (un)successful initiatives, and also, the results achieved and obstacles experienced by Grip op Water Altena and MMCO. Although both case studies of this research are examples of project-initiated and co-designed CBMs, a cross-case comparison of the themes, similarities, and differences across the two can still allow for generating valuable insights, and for providing such recommendations. On the one hand, the two CBMs had different thematic foci and this limited the possibility of a direct comparison of some of the aspects across cases. On the other hand, both CBMs were established under the Ground Truth 2.0 project and followed the same overall establishment methodology. These CBMs operate in fundamentally different contexts; one in a rural setting in a developing country in Africa and the other in an urban area in a developed country in Europe. This provided the possibility of a comparative analysis across the two cases. Similar to chapters 4 and 5, the CPI Framework was used as a frame for this comparative analysis. The results of this analysis are presented in sections 6.1 to 6.5 of this chapter. The chapter is concluded in section 6.6 with a reflection on the cross-case analysis of Grip op Water Altena and Maasai Mara Citizen Observatory.

---

[54] This chapter is partially based on: Gharesifard, M., Wehn, U., & van der Zaag, P. (2019a). Context matters: a baseline analysis of contextual realities for two community-based monitoring initiatives of water and environment in Europe and Africa. *Journal of Hydrology*, 124144. doi:https://doi.org/10.1016/j.jhydrol.2019.124144

6.1 GOAL & OBJECTIVES: GRIP OP WATER ALTENA VERSUS MAASAI MARA CITIZEN OBSERVATORY

A summary of the cross case comparison of Grip op Water Altena and MMCO across the Goals and Objectives dimension and aspects of the CPI Framework is provided in Table 6.1.

Regardless of their thematic focus, the vision and mission of both MMCO and Grip op Water Altena portray CBMs that aim at supporting environmental stewardship. In addition, Grip op Water Altena aims at facilitating cooperative planning for management of pluvial flooding in Land van Heudsen en Altena, while MMCO aims at contributing to collection and sharing of biodiversity and livelihood-related data in the Mara ecosystem. Both CBMs are operating in fundamentally different institutional systems; one within a system that has constitutional and legislative provisions for public participation in environmental management and the other one within a highly institutionalized system of water management with little or no provision for public participation.

The overarching objectives of the two CBMs were defined using a consistent approach for co-design, in consultation with representatives of different groups of stakeholders in each case. In both cases the overarching objectives were defined very broadly and with the aim of being as inclusive as possible. Therefore, in both CBMs, the co-designed objectives align well with the identified actor-specific goals. Nevertheless, this striving for accommodating diverse wishes resulted in overly ambitious aims for Grip op Water Altena and MMCO; ambitions that were only partly achieved by the end of the Ground Truth 2.0 project. While MMCO did not experience a change in objectives, the attention of Grip op Water Altena slowly changed towards two specific objectives. One of these objectives was setting up a knowledge platform for exchanging perspectives and tips to take measures against damage from pluvial flooding, and the other one was supporting open and constructive dialogue between citizens, the Regional Water Authority Rivierenland and municipality of Altena. This change in focus was influenced by existing contextual settings in the Dutch case, namely a highly institutionalized system of water management with well-established processes for communication and information exchange and little interest by the authorities for changing these practices.

In both cases, there was no formal procedure in place for monitoring the achievement of the objectives during the establishment process. Partial achievement of the objectives in Grip op Water Altena was mainly result of a long process of establishment and consensus building (also linked to the fact that the GT2.0 co-design methodology was being developed at the same time), a lack of sense of urgency of the topic within the community that is linked to the identified awareness gap and high trust of Dutch citizen in authorities for keeping them safe from floods, and also the absence of pluvial floods since the CBM was established. In the case of MMCO, several factor contributed to the partial achievement of the objectives e.g. lack of sense of local ownership, social and technological constraints, the issue of data sensitivity and linked to that the time-demanding process of consensus building for developing a data sharing policy, as well as a low level of engagement and actual use of tools.

Table 6.1 Summary of the cross case comparison across the Goals and Objectives dimension and aspects of the CPI Framework

Dimension	Aspect	Grip op Water Altena	MMCO

Goals & objectives	Overarching objectives	• Higher aim: Facilitating cooperative planning and environmental stewardship around the topic of pluvial flooding in Altena
		• The focus is more on the third and forth objectives i.e. Setting up a knowledge platform for exchanging perspectives and tips to take measures against damage from pluvial flooding, and supporting open and constructive dialogue between all involved stakeholders.
		• Reasons for shortcomings:
		  o Long establishment process
		  o Lack of awareness and sense of urgency of the topic within the community
		    o Highly institutionalized system of water governance in the Netherlands
			• Higher aim: Promoting environmental monitoring and environmental stewardship with the aim of balancing sustainable livelihoods and biodiversity management in the Mara
			• No change in focus; the objective was development of a socio-technical platform that contributes to increased access to data and information, and support better decision and policy making, which in turn lead to a better balance between biodiversity conservation and livelihood management
			• Reasons for shortcomings:
			  o Low level of engagement and actual use of tools
			    o Lack of sense of local ownership
			    o Low level of internal communication within member organizations
			      o Conflicting interests
			      o Cultural norms
			      o Social and technological constraints
			      o data sensitivity issues

	Goals for different actors	• The co-designed objectives align very well with the actor-specific goals of all stakeholders involved; namely, the municipality, water authorities and local community members.	• The co-designed objectives align with interests and wishes of many stakeholders, but this large scope was interpreted as producing generic and un-structured data, with a low level of usefulness for different organizations

	Monitoring objectives	• No formal procedure for monitoring the objectives, except for:
		  o Revisiting, and informally taking stock of the objectives during collaborative meetings
		  o Reverse impact journey
			• No formal procedure for monitoring the objectives, except for:
			  o Revisiting, and informally taking stock of the objectives during collaborative meetings
			  o Reverse impact journey

	Change of objectives over time	• A change in focus of the objectives that resulted in less focus on collecting and sharing data about weather and water systems and also supporting short communication lines between different stakeholders. • The change in focus was influenced by the authorities and the heavily institutionalized water management practices in the Netherlands.	• The objectives were defined quite broadly and there were no changes in the agreed-upon objectives

6.2 PARTICIPATION DYNAMICS: GRIP OP WATER ALTENA VERSUS MAASAI MARA CITIZEN OBSERVATORY

In terms of geographic scope, Grip op Water Altena is much smaller than MMCO. In comparison, the total surface area of the Dutch case study is less than 3% of the Kenyan one and it has less than 7% of the population of the Kenyan case. Moreover, the population in the Dutch case is mainly comprised of urban-dwellers living in towns, as opposed to the community members in the Kenyan case that are mainly geographically dispersed rural inhabitants.

The baseline analysis of the two cases revealed that despite the legislative differences, the role of the general public in relevant decision making processes in both cases was minimal, often limited to electing representatives. The co-design process in both cases provided an opportunity for a number of key stakeholders to come together and jointly co-design the functionalities of each CBM. In case of Grip op Water, the Regional Water Authority Rivierenland, Municipality of Altena, a small number of interested community members (often of higher average age ranges, and mainly male) and a few NGOs were the main participants in the co-design process. Participants in the co-design processes of MMCO included representatives of local and national level organizations (e.g. the Narok County Government, KWS, KFS, etc.), universities (Egerton University and Maasai Mara University), schools, NGOs (e.g. the African Conservation Centre and Friends of Maasai Mara and MMWCA), as well as a small number of individual community members and representatives of organized community groups.

Both CBMs had difficulties with engaging a large number of members and end-users. In case of Grip op Water Altena, the exact number of end-users is unknown, but most probably will align to a great extent with the co-design group and include the water authorities, municipality and a small group of local community members who know about this CBM. Similarly the number of end-users of MMCO is also unknown, but schools, universities and the Meteorological Department of the Narok County are among the current end-users of MMCO tools.

Participation in Grip op Water requires access to internet and smart phone or computer, which is widely accessible in the Netherlands, as well as a fairly small time commitment (e.g. to exchange information on a website or travel a small distance to participate in a meeting). In case of MMCO, however, online participation requires access to internet and smart phone, which are not widely accessible in the case study area. Moreover, the time spend for data collection and sharing and the financial burdens for online and offline participation, e.g. the cost of data bundle, commute expenses for data collection (e.g. fuel), or costs of accommodation and meals for participation in a face-to-face meetings are very high. Moreover, using MMCO's Apps requires a basic training.

Based on the results of the baseline analysis in the two cases, the communication paradigm about the issues in focus of each case was mainly uni-directional from authorities to citizens. Both CBMs facilitated communication and information exchange between different stakeholders by creating a number of two-way channels of communication among the stakeholders. However, in both cases interactive dialogues mainly happened during the face-to-face meetings. Both cases formed WhatsApp groups and used this as an online channel for interactive communication among the stakeholders; however, MMCO's WhatsApp group has many more members and it is more active, as compared to the WhatsApp group in Grip op Water Altena.

Table 6.2 depicts the cross case comparison of Grip op Water Altena and MMCO across the Participation dimension and aspects of the CPI Framework.

Table 6.2 Summary of the cross case comparison across the Participation dimension and aspects of the CPI Framework

Dimension	Aspect	Grip op Water Altena	MMCO
Participation	Type of initiative	• Cooperative planning and environmental stewardship • Controversial on the environmental stewardship domain	• Environmental monitoring and environmental stewardship • Constitutional and legislative provisions allow for participation of all stakeholders
	Geographic scope	Land van Heudsen en Altena, which is located at Municipality of Altena. • 211.75 km2 • 55840 inhabitants	Mara ecosystem on the Kenyan side • 7,921.2 km2 • 850920 inhabitants (geographically dispersed)
	(non)participant groups	• Participants in the co-design processes: o A small number of local community members (higher average age ranges, and mainly male) o Environmental NGOs o Municipality of Altena o The Regional Water Authority Rivierenland • Several national and provincial level actors (e.g. Rijkswaterstaat, Ministry of Infrastructure and Water Management, Union of Regional Water Authorities, the Association of Dutch municipalities, and the Province of Noord-Brabant) were not involved • Number of end-users: unknown, but most probably water authorities, municipality and a small group of local community members who know about this CBM	• Participants in the co-design processes: o Representatives of local and national level organizations (e.g. the Narok County Government, KWS, KFS, etc.) o Egerton University and Maasai Mara University o NGOs (e.g. the African Conservation Centre and Friends of Maasai Mara and MMWCA o Individual community members and representatives of organized community groups • Low representation of local community members and private tourism sector. • National level stakeholders such as the Ministry of Environment and Forestry, Ministry of Agriculture, Livestock, Fisheries and Irrigation, and Kenya Market Trust were not represented. • Number of end users: unknown, but include schools, universities and the Meteorological Department of the Narok County.
	Efforts required to participate	Online: • Internet, smart phone or computer, time to exchange information or promote the activities of the CBM via social media Offline: • Time commitment • Travel expenses	Online: • Internet, smart phone, time spend for data collection and sharing, financial means (data bundle and commute costs for data collection, e.g. fuel), training Offline: • Time commitment • Financial means (incl. travel and accommodation and food expenses)
	Support offered for participation	• Organizational support • Technical support • Financial support	• Organizational support • Technical support • Financial support • Capacity building
	Patterns of communication	• Created a number of two-way communication possibilities among the stakeholders • An online mechanism for interactive dialogue is lacking • Interactive dialogue is mainly limited to the offline mode and via face-to-face interactions	• Created a number of two-way communication and information sharing possibilities among the stakeholders • Online interactive dialogue and information sharing using a dedicated WhatsApp group • Interactive dialogue via face-to-face interactions
	Communication and participation method	• Change in communication modes by facilitating awareness raising, communication, and data & information sharing between different stakeholders. • No change in decision modes	• Change in communication modes by facilitating awareness raising, communication, and data & information sharing between different stakeholders. • No change in decision modes

6.3 POWER DYNAMICS: GRIP OP WATER ALTENA VERSUS MAASAI MARA CITIZEN OBSERVATORY

Regardless of the fact that Grip op Water Altena and MMCO are co-designed CBMs, they were both project-driven initiatives and therefore certain project requirements, interests and initial framings influenced the objectives, establishment process, functioning and subsequently results of these initiatives. While there was an overall agreement that Grip op Water Altena was co-designed (perhaps with more influence from the GT2.0 team and authorities), there was no agreement about the establishment process of MMCO; an aspect that may indicate that internal power dynamics in this CBM are seen differently by different individuals.

Both CBMs increased access to data for local stakeholders, nevertheless, the extent of this change was very limited in both cases. Grip op Water Altena produced very little new data and the integrated data and information on the web-platform of this CBM were mostly accessible through other sources. In case of MMCO, most frequently mentioned increase in access to data was related to the data from weather stations and not to collected observations via the CBM Apps, as these were very few. Moreover, sensitivity of some data and information produced in MMCO resulted in restricting the access to and control over that part of the data and information produced in this CBM to government organizations.

None of the two CBMs contributed to a change in the level of authority and power of different stakeholders; nevertheless, they both provided the participants with possibilities for exerting communicative influence on decision making processes. In case of Grip op Water Altena, this was made possible by providing alternative possibilities for dialogues and information exchange about the issue of pluvial flooding among the stakeholders. MMCO contributed to reinforcing already existing possibilities for interactions among stakeholders, as well as by increasing knowledge and awareness about the topics of biodiversity conservation and livelihood management among its members.

Regarding future revenue streams; both CBMs are planning to use a 'sponsorship' model, in which a government organization will provide the financial resources required to sustain the CBM. At the time of conducting this research, it was decided that funding required for continuation of Grip op Water Altena activities will be mainly provided by the municipality and water authority. However there was no agreement or official commitment from CBM members in MMCO in this regard. Table 6.3 provides an overview of the Power dynamics dimension and aspects of the CPI Framework for the two cases.

Table 6.3 Summary of the cross case comparison across the Power Dynamics dimension and aspects

Dimension	Aspect	Grip op Water Altena	MMCO
Power dynamics	Social/institutional & political context	• Merger of the municipalities • 'Subsidy Klimaatactief' • Additional structural measures, which make the situation look better and may lead to reducing the sense of urgency and less attention to the topic from the community members • No recent flood incident. This has decreased attention and sense of urgency of the topic among local community members	• No major direct change because of the general election of 2017, but staff turnover in some ministers and departments within the Narok County Government negatively affected the efforts for engaging this stakeholder. • More restricted access to the national reserve, especially for research purposes, that can negatively affect data collection in the park
	Establishment mechanism	• Project-driven • The majority agree that this CBM was co-designed, but a few believe that the authorities and the GT2.0 team members had more influence than community members	• Project-driven • Disagreement on the establishment process of MMCO shows that there is no common understanding about the establishment process of this CBM and indicates that internal power dynamics in this case are perceived differently by different individuals.

	Access to and control over data	• Change in access to data is perceived differently by the GT2.0 team members and the CBM members
		  ○ GT2.0 team: Change in access due to integration of various information
		  ○ CBM members: Little or no change in access; available information was accessible through other sources
			• Agreement about limited change in access to data among stakeholder groups
			• A data sharing policy is designed for determining the level of access to data, but this policy is not (yet) internalized by the stakeholders and there are disagreements on its main principles

	Authority and power	• No change in the level of authority and power of different stakeholders • Alternative possibility for communicative influence	• No change in the level of authority and power of different stakeholders, • Reinforced already existing communicative influence possibilities by increasing interactions, as well as participants' knowledge and awareness
	Revenue stream	• Agreed upon revenue streams, mainly through government sponsorship and organizational support by ANV	• Government sponsorship model is the most probable option, but future revenue streams are (yet) unclear and there is no agreement or official commitment from CBM member organizations to provide this financial support.

157

6.4 TECHNOLOGICAL CHOICES: GRIP OP WATER ALTENA VERSUS MAASAI MARA CITIZEN OBSERVATORY

There are quite a few differences in the technological choices in Grip op Water Altena and MMCO. The main technological component of Grip op Water Altena is a web-platform that supports access to, and possibilities for sharing water and weather-related information, while MMCO uses two Apps, a number of physical sensors and a web-platform that enable collection and sharing of data and information about the topics of biodiversity conservation and sustainable management of livelihoods in the Mara.

The baseline analysis of access to technology in the two case studies of this research presented a classic North-South or 'developed versus developing' country situation. This baseline analysis demonstrated a drastic difference between the Netherlands and Kenya in terms of both, diffusion and actual use of ICTs; a difference that was amplified by the low level of literacy and required skills for effective use of ICTs in the Kenyan context. Based on to this baseline situation, the technological choices of Grip op Water Altena were evaluated as highly accessible, user friendly and fit for the local context (e.g. in terms of the local language used). On the contrary, members of MMCO believe that technological choices in this CBM are not accessible enough for average local community members, due to the aforementioned social and technological constraints and also because of a number of design choices, such as developing text-based Apps in the English language and not in the local language. All in all, a combination of social and technological context and the design choices in the two initiatives produced quite an inclusive CBM in the Dutch case study, while it resulted in exclusion of a number of groups from MMCO (see Table 6.4).

*Table 6.4 Summary of the cross case comparison across the Technology dimension and
aspects*

Dimension	Aspect	Grip op Water Altena	MMCO
Technology	Technologies used	A web-platform that enables: • Access to, and possibilities for sharing water and weather-related information • Communication channels are mainly external or offline	Two Apps, a number physical sensors and a web-platform that enable: • Collection and sharing information about the topics of biodiversity conservation and sustainable management of livelihoods in the Mara • Communication channels are mainly external or offline
	Accessibility of the technologies used	Agreement on accessibility of technologies • The web-platform is very accessible and easy to use • Use of some of the maps may require expertise or knowledge	Disagreement about accessibility of technologies between GT2.0 team members and CBM members • GT2.0 team members believe technologies are fairly accessible for a large part of the population • CBM members believed that technologies are not accessible enough for an average local community member because of: 　o Limited access to smart phones and internet 　o Low average level of literacy and digital skills 　o The need for an initial training 　o The text-based design of the App 　o Design in English language 　o Costs of data bundles
	Included/excluded groups	• No major groups excluded because of: 　o User-friendly web-platform 　o Use of visual material 　o Use of local language 　o Form-based observation submission options	• Quite a few groups are excluded, e.g.: 　o People without a smart phone or the technical skills required to use one 　o People with non-Android smart phones 　o Non-English speaking local community members 　o Illiterate people

6.5 RESULTS: GRIP OP WATER ALTENA VERSUS MAASAI MARA CITIZEN OBSERVATORY

Grip op Water Altena, provided offline and online possibilities for interaction and information exchange between stakeholders. Information sharing via the web-platform was mainly one directional and from authorities to citizens. Citizens also occasionally share information via this web-platform, but at the time of conducting this research this was very limited. In comparison, the most tangible realized outputs of MMCO were its technological components that included the two Apps, the website, and the weather stations and screens. Added components from existing tools and services, the WhatsApp group of MMCO, produced data and information in this CBM and the educational material related to mapathons and weather stations were also among the elicited realized outputs in this initiative. Moreover, MMCO has a slight chance of having a potential negative impact that relates to collection and sharing of sensitive biodiversity data in this CBM (i.e. due to a potential a leak of data and information produced in this CBM). Nevertheless, a data sharing policy is envisaged for eliminating or reducing this negative output.

In terms of realized outcomes, MMCO and Grip op Water Altena were very similar. Both CBMs created a (small) community of stakeholders around the issue in their focus. Awareness raising about participatory approaches for environmental management was among the realized outcomes of both initiatives, however, in the case of MMCO, this awareness raising also extended to highlighting the existing data gaps related to biodiversity conservation and sustainable management of livelihoods. Moreover, both initiatives succeeded in creating new forms of offline and online communication and interaction between different stakeholders.

Future outcomes and impacts in both cases are largely uncertain and to a great extent depend on their future sustainability and uptake. Table 6.5 provides the details of identified expected future outcomes and impacts by interviewees for the two CBMs. A comparison between the expected categories of outcomes and impacts of the two CBMs shows that while both initiatives have the potential to yield a wide range of individual, societal, environmental and governance-related changes, the range of expected changes for MMCO is more diverse and includes expected changes for science and economy. This is linked to both existing data gaps in the Kenyan case (as compared to the Dutch case), as well as the thematic focus of this case. More specifically, while the issue in focus of Grip op Water Altena is not perceived as urgent by many Dutch citizens, MMCO's thematic focus is perceived as both important and urgent because of its direct link to livelihoods of the community members and the economy of the Mara region.

Table 6.5 Summary of the cross case comparison across the Results dimension and aspects

Dimension	Aspect	Grip op Water Altena	MMCO
Results	Outputs	• Realized outputs: 　o Online and offline possibilities for interaction and information exchange between stakeholders via a web-platform and face-to-face meetings 　o One directional online information sharing from authorities to citizens and other stakeholders 　o Limited and occasional online information sharing from citizens to other stakeholders	• Realized outputs: 　o Technological components including the two Apps, the website, weather stations and screens 　o Integrated components from existing tools and services 　o MMCO WhatsApp group 　o Produced data and information 　o Educational material related to mapathons and weather stations 　o Production of sensitive biodiversity data
	Outcomes	• Realized outcomes: 　o Created a (small) community of stakeholders around the topic of pluvial flooding 　o Awareness raising about participatory approaches for reducing the risk of pluvial flooding 　o Creating new forms of communication and interaction between different stakeholders • Expected future outcome: No major change was expected to take place in the near future because of Grip op Water Altena, but this CBM has the potential to contribute to more awareness raising, data sharing, and better communication and interaction among the stakeholders.	• Realized outcomes: 　o Knowledge sharing and awareness raising (at a small scale) about the concept of community-based monitoring 　o Creating knowledge and awareness about existing data gaps 　o Creating a (small) community of stakeholders around the topic of biodiversity conservation and sustainable management of livelihoods 　o Creating new forms of communication and interaction between different stakeholders • Expected future outcomes: Change in attitude towards important issues such as conservation, communication and exchange of data and information between different stakeholders, generation of scientific outcomes and contribution to fact-based decision making that can in turn result in economic changes for individuals and communities.
	Impacts	• Future impact (depending on future uptake): 　o Governance-related impacts including contribution to a change in policy and moving towards a participatory approaches for water management, as well as increased awareness, trust and transparency among stakeholders 　o Societal impacts including contribution to a change in attitude towards environmental stewardship 　o Environmental impacts such as contribution to creating a more flood-proof Altena	• Future impact (depending on future uptake): 　o Governance-related impacts including contribution to fact-based environmental decision and policy making and conservation actions, engaging community member and giving them a voice in decision making processes, planning for uncertainties such as climate change, and facilitating more transparent and accountable environmental governance. 　o Economic impacts for both individual and societal levels 　o Environmental impacts such as avoiding overgrazing, reforestation, reducing human-wildlife conflict and conservation of biodiversity and other natural resources

6.6 CONCLUSIONS

This chapter presented the results of a cross case analysis that was conducted to compare and contrast the baseline situation, establishment process and results of the two case studies of this research. Due to the fact that Grip op Water Altena and MMCO have different thematic foci, a direct comparison of some of the factors that affect the establishment and functioning of each initiative may not be meaningful, nevertheless, this cross case comparison yielded interesting results that are summarized hereafter.

Due to the fact that both CBMs were established using a consistent co-design approach and through a process of consensus making among diverse groups of stakeholders, they both ended up with quite broad and ambitious visions, missions, and objectives that accommodate diverse stakeholders' wishes. This increased the alignment of the CBM objectives with actor-specific goals, and at the same time, made it very difficult to achieve these objectives within the three year timeframe of the Ground Truth2.0 project. Moreover, both initiatives were focusing on creating a social change towards environmental stewardship and Grip op Water Altena also aimed at facilitating an institutional change towards cooperative planning regarding the issue of pluvial flooding in Altena. Both of these higher aims are ambitious and require creating an interactive dialogue between citizens and decision makers (Wehn et al., 2015a), which is a time demanding process that contributed to the partial achievement of the objectives in both CBMs so far.

Regardless of the fact that the two CBMs have different geographic foci and a very different pool of potential participants, engagement of a large number of CBM members and end-users proved to be a challenge for both cases. Nevertheless, both cases succeeded in facilitating communication, information exchange and dialogue among small groups of stakeholders. One of the determining factors for future use of the tools in CBMs is efforts required, and support offered, for using those tools. The results of this cross case comparison shows that using the tools developed in Grip op Water Altena requires little resources and efforts, while the use of MMCO is hindered by multiple social and technological contextual settings (e.g. low level of access to technology, literacy and digital skills), and a number of design choices (e.g. the text-based design of the Apps in English).

The initial need for establishing both CBMs was project-driven and they followed a consistent co-design methodology. Nevertheless, power dynamics during the co-design process were perceived differently by members of Grip op Water Altena and MMCO. There was an overall consensus that Grip op Water Altena was in fact co-designed, while the establishment process of MMCO was perceived differently by different individuals. Shift in power dynamics because of participation in both CBMs mainly relates to communication and connectivity among stakeholders. Nevertheless, both CBMs failed to change and improve access to or control over data and information.

In terms of technological choices, a combination of social and technological context of the two cases and the design choices made in Grip op Water Altena and MMCO resulted in quite an inclusive CBM in the Dutch case study, while resulted in exclusion of quite a few groups from MMCO.

Finally, the results of this cross case comparison depicted several outputs across the two cases and showed similarities in terms of (lack of) resized outcomes of the two CBMs. More specifically, both Grip op Water Altena and MMCO contributed to creating a small community and raising awareness about a common environmental challenge. Nevertheless, future contribution of MMCO and Grip op Water Altena to solving complex environmental challenges such as pluvial floods in Altena and balancing sustainable livelihoods and biodiversity conservations in the Mara are largely uncertain. Nevertheless, in order to create further changes or have an impact, both CBMs need to maintain their activities and current members, and strive for engaging more active members.

7

CONCLUSIONS

The main objective of this research was to conduct a systematic evaluation of the factors that influence the establishment, functioning and outcomes of two CBM initiatives. Based on the findings of this research that were presented in Chapters 4, 5 and 6, section 7.1 summarizes the main conclusions of this research. Section 7.2 presents the major contributions of the research, both in terms of conceptualization of CBM initiatives and the knowledge generated from closely studying the process of establishment and functioning of two CBMs. Reflections on the methodological approach of the research are outlined in section 7.3. Section 7.4 highlights the limitations of the research. Finally, in line with the fourth objective of this research (i.e. providing recommendations), a number of recommendations are proposed in section 7.5 that can have applications for existing and forthcoming CBMs.

7.1 MAIN CONCLUSIONS OF THE RESEARCH

This research aimed at conducting a systematic evaluation of the factors that influence the establishment, functioning and results of CBM initiatives. The findings of the qualitative empirical research presented in this study demonstrate that systematic evaluation of goals and objectives, participation processes, power dynamics, technological choices and results of CBMs can provide critical insights into the establishment process and functioning of these initiatives. Moreover, it can be concluded that factors influencing the establishment process and functioning of CBMs are not only internal to the initiatives, but also context-related. The interplay between internal and context-related factors provides an interpretation of the notion of community in a particular CBM; a concept which is otherwise difficult to depict.

The results of a two phase empirical research into the factors that influenced the establishment and functioning of two co-designed CBMs in the context of Ground Truth 2.0 project yielded twelve specific conclusions that are presented hereafter. Due to the fact that the two CBMs studied in this research were both project-initiated and co-designed, generalizability of some of the following conclusions may be limited. Nevertheless, this issue is carefully considered and, where relevant, acknowledged in the following conclusions.

- Without a sound understanding of the contextual realities in which a CBM will operate, assumptions need to be made in order to understand what can be expected from that initiative. Moreover, there will be no baseline for measuring outcomes and impacts of the initiatives. Therefore, it is highly important that such a baseline analysis be conducted at an early stage and preferably before the establishment of a CBM initiative.

- CBMs should strive for realistic and specific objectives and carefully consider actor-specific goals and contextual settings (e.g. technological infrastructure and existing power relationships between stakeholders) in defining their objectives. This is especially important in the case of CBMs that aim at moving beyond the environmental monitoring function and engage with policy and decision making processes.

- Perceived importance or urgency of the topic, existing power relationships, level of trust among the actors, length of the establishment process and ease or difficulty of participation are all among the factors that affect the initial and continued participation of stakeholders in a CBM.

- Establishing CBMs using a co-design approach is a time-demanding and resource-intensive process that requires efforts and commitment from all involved actors. CBMs that follow a co-design methodology should set a clear timeframe for defining their aims, objectives and functionalities and participants in the co-design process should be made aware of the time commitment they need to make for participation.

- Co-design process provides possibilities for discussion and consensus building among different stakeholders and thus provides a more equal chance for parties involved to influence the establishment processes of a CBM. Nevertheless, the fact that a CBM is co-created or co-designed does not mean that power relationships between stakeholders do not exist or are balanced out completely. Existing power relationship among the actors, the issue of data ownership, source of technical, financial and organizational support for the CBM, and interests of actors involved in the establishment process are among the factors that help with understanding and reflecting on power dynamics in a CBM.

- Technological components of CBMs are shaped by wishes and preferences of different actors involved in their design and may reshape or maintain certain power relationships between those actors.

- Establishing ICT-mediated CBMs in developing countries and regions with less technological advancements is particularly more challenging and requires careful considerations for inclusion of vulnerable and less tech-savvy community members.

- Compatibility of technological choices with social, institutional and technological context reduces the chance of excluding major group within society. Nevertheless, heterogeneity of society should be acknowledged and realistic expectations should be set and communicated about the extent to which CBMs can enable participation of different groups within society. Perhaps more broadly: sometimes progressive choices may need to be made to include and support certain groups and not others.

- Having a clear picture of the baseline situation in which a CBM is being established and distinguishing between outputs, outcomes and impacts facilitates studying the results of a CBM.

- Data, information and knowledge exchange, awareness raising, learning opportunities, and communication and interaction possibilities created because of a CBM are among the more immediate, tangible and easier to study results of CBMs. In contrast, environmental impacts and shift in power-relationships among stakeholders are long-term, maybe slow-moving and therefore more difficult to study.

- The extent to which project-driven CBMs, with no official mandate, can contribute to solving complex environmental challenges or balancing existing and un-even power relationships is often limited.

- Establishment of project-driven CBMs is likely to be influenced by factors such as pre-framing of challenges and possible solutions, existing interests and expertise within initiating organizations, as well as pre-defined resources, time-frame and other obligations towards funding organizations. The need for establishing such initiatives and their functionalities may therefore be 'supply-driven' and not 'demand-driven'; an issue that can undermine engagement efforts by these initiatives and negatively influence their sustainability.

7.2 MAJOR CONTRIBUTIONS OF THE RESEARCH

This research has made two major contributions. Firstly, this research contributed to the conceptualization of CBMs by developing a multi-dimensional conceptual framework that is suitable for context analysis, process evaluation and impact assessment of CBM initiatives. This conceptualization provides an interpretation of what 'community' means in the context of a CBM initiative. The first contribution of this study is detailed in section 7.2.1. Secondly, the results of this research contributed the body of knowledge about the establishment, functioning and outcomes of CBMs in a number of ways that are explained in section 7.2.2.

7.2.1 Contribution to conceptualization of community-based monitoring initiatives

CBM initiatives have a high perceived potential for enhancing informed, inclusive, democratic and transparent environmental decision making. However, the conceptual understanding required to evaluate and critically review the dynamics at play that might enable or hinder these initiatives from delivering on their potential is limited. In line with this gap in knowledge and the ambitions of this study, the first research objective was set to develop a conceptual framework for evaluating the factors that influence the establishment, functioning and outcomes of CBM initiatives. Based on an extensive review of the literature and theories in the field of Citizen Science and other affiliated fields of research, the CPI Framework was developed that consists of five dimensions that can help unpack what influences the establishment and functioning of a CBM.

The CPI Framework has three different applications. Firstly, it can be utilized as a framework for analyzing the baseline situation of contextual factors (such as the social, political, institutional and technological setting) that can influence the establishment and functioning of a CBM initiative. If conducted before the establishment of a CBM, such a baseline contextual analysis can provide valuable insights for the design of the initiative as well as for stakeholder engagement activities. Secondly, this framework can be utilized for impact assessment purposes. The classifications provided for outputs, outcomes and impacts (Table 2.2) help evaluate the results generated by a CBM in a systematic and

168

logical way. The third application of the CPI Framework is in process evaluation. Answers to the core questions raised by this framework (e.g. who participates in the CBM initiative and how, and who does not? or who controls and influences the CBM initiative and how?) help enhance our understanding of the processes that led to the outputs, outcomes and impacts of the CBM and provide valuable insights about why and how positive, negative, intended and unintended results were generated. This multi-purpose nature is a unique feature of the CPI Framework that is not present in any other conceptualization of Citizen Science projects or CBMs. The five dimensions of the proposed framework are common across all CBMs and all dimensions and aspects need to be closely examined in order to gain a thorough understanding of an initiative.

What constitutes community in a CBM initiative is highly context-dependent, dynamic and hence difficult to analyze. In general terms, a community is a social unit consisting of a group of individuals who have something in common. In the context of a CBM initiative, this common denominator is usually interest in, concerns about or stake in an environmental issue. However, the community in the context of CBM initiatives does not exist as a clearly defined and static entity, rather it is shaped and reshaped at any point in time by factors such as the initiation of the CBM, the composition and the group of actors involved, their goals and interests, the power dynamics among the actors and enabling technologies. Studying CBMs using the CPI Framework enhances the understanding about several dynamic processes and aspects about the establishment and functioning of these initiatives, and hereby facilitates interpretation of the meaning of community in the context of a particular CBM.

7.2.2 Contribution to the body of knowledge about the establishment, functioning and outcomes of community-based monitoring initiatives

This research contributed to the body of knowledge about the establishment, functioning and outcomes of CBMs in two distinct ways that are detailed hereafter.

The 7th Framework Programme for Research and Technological Development in Europe (FP7) allocated more than 26 million Euros to five projects for demonstrating and testing the concept of Citizen Observatories in Europe (i.e. WeSenseIt, CITI-SENSE, Citclops, COBWEB and OMNISCIENTIS). Despite this investment and the efforts spent on establishing several CBMs in the context of aforementioned projects, lessons learned from setting up these initiatives and factors that affected their establishment and functioning have not so far been studied in-depth and in a coherent manner. As indicated in Chapter 2, the process of developing the CPI Framework involved a review of the lessons learned from the establishment process of the five aforementioned 'pioneer' or 'legacy' CBM projects in Europe. The first contribution of this study to existing knowledge about the establishment, functioning of CBMs is thus the review, and incorporation of, the lessons learned from the establishment of several CBMs within the context of these projects in development of the CPI Framework.

The second contribution of this research to existing knowledge about the establishment process and functioning of CBMs is the results of testing the CPI Framework for studying two real life CBM initiatives throughout the lifetime of an EU-funded Horizon 2020 project. The study of the baseline situation, establishment process and results of Grip op Water Altena and MMCO provided detailed insights about the factors that affected the establishment process, functioning and results of these initiatives. Selection of the two case studies of this research allowed for conducting a comparative analysis of the factors that influence the establishment and functioning of two CBMs, one in a developed country in Europe and the other one in a developing country in Africa. These insights that are presented in Chapters 4, 5 and 6 of this dissertation are built on detailed empirical and observational evidence from the studying the two CBMs over a three years time period.

7.3 REFLECTION ON THE RESEARCH DESIGN AND METHODOLOGY

Reflecting on the research design and methodology of this study resulted in a number of conclusions that are presented hereafter.

While having demonstrated the advantages of conducting an in-depth baseline analysis at the initiation of a CBM initiative, it should be acknowledged that undertaking such a thorough contextual analysis initiatives requires substantial time, resources and expertise, which may not always be available for all CBM initiatives or projects. Adopting the methodology developed for this research requires a detailed context mapping as well as close observations of every step in establishing a CBM, and involves developing interview protocols, setting up and conducting interviews, interpreting the results. Nevertheless, a less comprehensive context mapping that is tailored to available resources and based on a desk research of secondary data sources can still yield valuable insights and is therefore highly recommended for all CBM initiatives and projects.

Moreover, achieving the third research objective of this study, and linked to that answering its sixth research question, required evaluation of evolving processes, outputs and interim outcomes of the two CBMs over time (approximately three years). Although the two phase design of this research enabled studying quite a few evolving processes in the establishment of the two CBMs (e.g. change in the objectives, participation processes, power dynamics etc.), the timeline of this study did not allow for evaluating the evolving outputs and interim outcomes of the two CBMs. The second phase of the empirical research of this study was conducted in November 2019 (i.e. towards the end of the Ground Truth 2.0 project) and included assessment of outputs and interim outcomes of the two CBMs. Nevertheless, in order to capture the evolving outputs and outcomes, there is a need for revisiting the cases in the future (e.g. in one to five years); a requirement that is beyond the timeframe envisioned for a PhD research.

7.4 LIMITATIONS OF THE RESEARCH

- The first limitation of this research relates to language barriers. Although the Netherlands has the second highest English language proficiency rank in Europe (EF Education First, 2015) and the majority of Dutch citizens speak fluent English, Dutch was the main language used in the co-design meetings of Grip op Water Altena. Therefore, the researcher relied on his limited Dutch language skills and mainly on the Dutch speaker members of the Ground Truth 2.0 team and meeting notes to follow the conversations in these meetings. Moreover, the majority of the interviews in this case were conducted in Dutch. For this purpose, the researcher had to train other interviewers and rely on their interviewing skills. Nevertheless, instruction sessions were held to train those interviewers and familiarize them with the interview protocols and how to conduct the interviews. Language was also a barrier in the Kenyan case. Although English and Swahili are the two official languages in Kenya, some local community members (especially elderly) in the Mara only speak KiSwahili. Interviews in this case were conducted in English and therefore excluded the non-English speaking local community members.

- The next limitation is closely linked to the timeline of conducting this study and its ambitions. As explained in the reflection on the research design and methodology, the third objective of this research called for evaluating the evolving processes, outputs and interim outcomes of the two CBMs over time. The establishment process of Grip op Water Altena and MMCO was a time-demanding process that started in early 2017 and their outputs and outcomes took quite some time to materialize. As indicated earlier, this was partly due to the fact that the Ground Truth 2.0 was developing the co-design methodology in parallel with establishing the CBMs In order to measure the outputs and outcomes of the two CBMs, the researcher had to conduct the second phase of data collection and analysis as close as possible to the end of the Ground Truth 2.0 project. The second phase of data collection was therefore only conducted in November 2019. By then, this doctoral research was already in its fifth year and this timeline did not allow for revisiting and measuring the evolving outputs and outcomes of Grip op Water Altena and MMCO another time. Therefore, the third objective of this research and its sixth research question are only partly addressed.

- The third limitation relates to the fact that both CBMs that were studied in this research were project-based and were established using a consistent co-design methodology. This may have limited the generalizability of some of the results of this research for top-down or bottom-up CBMs and also CBMS that are established using a different co-design approach.

7.5 RECOMMENDATIONS

In line with the fourth objective of this study, i.e. to provide recommendations for CBMs, this section presents a number of suggestions and considerations for establishing CBM initiatives. These recommendations are based on the insights generated from detailed analysis of Grip op Water Altena and MMCO and can help facilitate the establishment process and functioning of future and ongoing CBMs. Moreover, two topics of future research are identified and recommended that can help broaden the insights generated from this study.

7.5.1 Recommendations for CBMs

This research demonstrated that the context in which CBMs are being established matters and can affect the establishment process and functioning of a CBM. Moreover, conducting a baseline analysis before establishing a CBM facilitates measuring its outputs, outcomes and impacts. Therefore, it is recommended that forthcoming CBMs conduct a baseline analysis of the social, technological and institutional contextual settings, with which they are going to interact. This baseline analysis should be conducted before or at the early stages of the establishment of the CBM. The CPI Framework provides a guiding frame for identifying the context-related factors that should be considered and the methodology developed in this study is an example of how such baseline analysis can be conducted. Nevertheless, if time, resources and expertise for a thorough context analysis are not available, a less comprehensive context mapping, based on the available resources (e.g. using a desk research of secondary data sources) is highly recommended for all forthcoming CBMs.

The next recommendation related to expectation from, and transparency about, the immediate, mid-term and long-term outcomes and impacts of CBMs. CBM is a relatively new concept and often before the establishment of a CBM, local stakeholders have not heard about this concept. Projects, organizations and facilitators who are involved in setting up CBMs are sometimes unclear about the expected future changes resulting from CBMs, or underestimate the time required for some expected changes to materialize. Clear and realistic expectations should be set by CBM initiators about the extent to which an initiative can contribute to solving complex environmental challenges and balancing existing un-even power relationships. Moreover, these expectations should be clearly communicated with local stakeholders. Failing to manage expectations and communicating these facts may lead to a sense of mistrust and disappointment in CBMs and undermine the engagement efforts and positive changes that these initiatives are able to contribute to.

The last recommendation relates to the time considerations for co-designing CBMs. Funders, project facilitators and local stakeholders should be made aware of the fact that even when a co-design methodology is readily available, deliberation and consensus building, e.g. for setting up the aims and objectives of a CBM, agreeing on its functionalities or designing and technological components, is a time-demanding process. Foreseeing a clear, yet flexible timeframe for different phases of the establishment process of a CBM and clear communications about this timeframe with the involved actors is therefore highly recommended, and will help avoid future disappointments.

7.5.2 Areas of further research

The first area of recommended future research is related to measuring the evolving outcomes and impacts resulting from the establishment of MMCO and Grip op Water Altena. Revisiting these CBMs in the future (e.g. in two to five years), may result in additional insights about their sustainability, functioning and evolving results. The current case study can then be used as a benchmark for measuring changes that these two CBMs may contribute to in the future.

Furthermore, applying the CPI Framework for context analysis, process evaluation and impact assessment of other CBMs (co-designed or otherwise) is expected to result in a better understanding of the factors that affect the establishment and functioning of CBMs. Therefore, researchers and practitioners are encouraged to apply and test the proposed framework of this research to evaluate the features and functioning of different CBM initiatives and to report on the implementation of the framework.

REFERENCES

Alfonso, L., Mazzoleni, M., Pfeiffer, E., & Wehn, U. (2017a). Ground Truth 2.0 project deliverable reports; Deliverable D1.5; Functional Design. Delft, the Netherlands.

Alfonso, L., Pfeiffer, E., & Mazzoleni, M. (2017b). Ground Truth 2.0 project deliverable reports; Deliverable D1.6; Management and tracking tool of user requirements. Delft, the Netherlands.

Arpaci, A., Aspuru, I., Bartonova, A., Broday, D., Castell, N., Cole-Hunter, T., . . . Svecova, V. (2016a). Evaluation of the performance of the case studies, CITI-SENSE project deliverable D2.4 Kjeller, Norway.

Arpaci, A., Aspuru, I., Bartonova, A., Castell, N., Cole-Hunter, T., Cowie, H., . . . Verheyden, W. (2016b). Empowerment potential evaluation, CITI-SENSE project deliverable D5.5 Kjeller, Norway.

Arpaci, A., Bartonova, A., Cada, K., Moreno Carranza, J., Castell, N., Cole-Hunter, T., . . . Verheyden, W. (2015). Methodology and protocol for citizens' empowerment, CITI-SENSE project deliverable D5.3 Kjeller, Norway.

Bandara, W., Furtmueller, E., Gorbacheva, E., Miskon, S., & Beekhuyzen, J. (2015). Achieving rigor in literature reviews: Insights from qualitative data analysis and tool-support. *Communications of the Association for Information Systems, 37,* 154-204.

Bartels, A., Eckstein, L., Waller, N., & Wiemann, D. (Eds.). (2017). *Postcolonial Justice*. Netherlands: Brill Publishers.

Bartonova, A., Rubio, A., García, I., & Aspuru, I. (2016). Part 1 Evaluation of the performance of the user cases: Public Places, CITI-SENSE project deliverable D3.4. Kjeller, Norway.

Bertot, J. C., Jaeger, P. T., Munson, S., & Glaisyer, T. (2010). Engaging the Public in Open Government: Social Media Technology and Policy for Government Transparency. *IEEE Computer, 43*(11), 53-59.

Bijker, W. E., Hughes, T. P., Pinch, T., & Douglas, D. G. (2012). *The social construction of technological systems: New directions in the sociology and history of technology*: MIT press.

Bio Innovation Service. (2018). Citizen science for environmental policy: development of an EU-wide inventory and analysis of selected practices. Final report for the European Commission, DG Environment under the contract 070203/2017/768879/ETU/ENV.A.3, in collaboration with Fundacion Ibercivis and The Natural History Museum, November 2018.

Birkin, L., & Goulson, D. (2015). Using citizen science to monitor pollination services. *Ecological Entomology*. doi:10.1111/een.12227

Boles, O. J. C., Shoemaker, A., Courtney Mustaphi, C. J., Petek, N., Ekblom, A., & Lane, P. J. (2019). Historical Ecologies of Pastoralist Overgrazing in Kenya: Long-Term Perspectives on Cause and Effect. *Human Ecology, 47*(3), 419-434. doi:10.1007/s10745-019-0072-9

Bonney, R., Ballard, H., Jordan, R., McCallie, E., Phillips, T., Shirk, J., & Wilderman, C. C. (2009a). Public Participation in Scientific Research: Defining the Field and Assessing Its Potential for Informal Science Education. A CAISE Inquiry Group Report. *Online Submission*.

Bonney, R., Cooper, C. B., Dickinson, J., Kelling, S., Phillips, T., Rosenberg, K. V., & Shirk, J. (2009b). Citizen science: a developing tool for expanding science knowledge and scientific literacy. *BioScience, 59*(11), 977-984.

Bonney, R., Phillips, T. B., Ballard, H. L., & Enck, J. W. (2015). Can citizen science enhance public understanding of science? *Public Understanding of Science, 25(1)*, 2-16. doi:http://dx.doi.org/10.1177/0963662515607406

Bonney, R., Shirk, J. L., Phillips, T. B., Wiggins, A., Ballard, H. L., Miller-Rushing, A. J., & Parrish, J. K. (2014). Next steps for citizen science. *Science, 343*(6178), 1436-1437.

Bradford, R., O'Sullivan, J., van der Craats, I., Krywkow, J., Rotko, P., Aaltonen, J., . . . Schelfaut, K. (2012). Risk perception–issues for flood management in Europe. *Natural Hazards and Earth System Science, 12*(7), 2299-2309. doi:http://dx.doi.org/10.5194/nhess-12-2299-2012

Brännström, I. (2012). Gender and digital divide 2000–2008 in two low-income economies in Sub-Saharan Africa: Kenya and Somalia in official statistics. *Government Information Quarterly, 29*(1), 60-67. doi:https://doi.org/10.1016/j.giq.2011.03.004

Brossard, D., Lewenstein, B., & Bonney, R. (2005). Scientific knowledge and attitude change: The impact of a citizen science project. *International Journal of Science Education, 27*(9), 1099-1121.

Ciravegna, F., Huwald, H., Lanfranchi, V., & Wehn de Montalvo, U. (2013). Citizen observatories: the WeSenseIt vision, *Presentation at the INSPIRE Conference 2013*. Florence, Italy, 23-27 June.

Cleaver, F. (1999). Paradoxes of Participation: Questioning Participatory Approaches to Development. *Journal of International Development, 11*(4), 597-612.

COBWEB Consortium. (2015a). Policy brief - The Potential of Crowdsourced Environmental Data to Support Policy Making, COBWEB project deliverable D2.4. Edinburgh, UK.

COBWEB Consortium. (2015b). Value adding to crowdsourced data for decision making, COBWEB project deliverable D2.2. Edinburgh, UK.

COBWEB Consortium. (2016). Educational Aspects of COBWEB – final report, COBWEB project deliverable D9.5. Edinburgh, UK.

COBWEB Consortium. (2017). Final publishable summary report, COBWEB project deliverable. Edinburgh, UK.

Conrad, C. C., & Hilchey, K. G. (2011). A review of citizen science and community-based environmental monitoring: issues and opportunities. *Environmental Monitoring and Assessment, 176*(1-4), 273-291.

Conrad, C. T., & Daoust, T. (2008). Community-based monitoring frameworks: Increasing the effectiveness of environmental stewardship. *Environmental Management, 41*(3), 358-366.

Cooper, C., Dickinson, J., Phillips, T., & Bonney, R. (2007). Citizen science as a tool for conservation in residential ecosystems. *Ecology and Society, 12*(2).

Cortes Arevalo, V. J. (2016). *Use of volunteers' information to support proactive inspection of hydraulic structures.* (PhD dissertation), Delft University of Technology, Delft, the Netherlands.

Crall, A. W., Newman, G. J., Jarnevich, C. S., Stohlgren, T. J., Waller, D. M., & Graham, J. (2010). Improving and integrating data on invasive species collected by citizen scientists. *Biological Invasions, 12*(10), 3419-3428. doi:10.1007/s10530-010-9740-9

Davenport, S. (2013). Prioritise citizen feedback to improve aid effectiveness. Retrieved from http://www.theguardian.com/global-development-professionals-network/2013/jul/22/feedback-loops-citizen-development

Dickinson, J. L., Shirk, J., Bonter, D., Bonney, R., Crain, R. L., Martin, J., . . . Purcell, K. (2012). The current state of citizen science as a tool for ecological research and public engagement. *Frontiers in Ecology and the Environment, 10*(6), 291-297. doi:http://dx.doi.org/10.1890/110236

Dickinson, J. L., Zuckerberg, B., & Bonter, D. N. (2010). Citizen science as an ecological research tool: challenges and benefits. *Annual review of ecology, evolution, and systematics, 41*, 149-172.

EASME. (2016). *Have you heard about the concept of Citizens' Observatories?* : the European Commission, Retrieved from https://ec.europa.eu/easme/en/news/have-you-heard-about-concept-citizens-observatories.

EF Education First. (2015). EF EPI: EF English Proficiency Index. http://www.ef.nl/epi/downloads/

Eleta, I., Clavell, G. G., Righi, V., & Balestrini, M. (2019). The Promise of Participation and Decision-Making Power in Citizen Science. *Citizen Science: Theory and Practice, 4*(1).

Emmett Environmental Law and Policy Clinic. (2019). A manual for citizen scientists starting or participating in data collection and environmental monitoring projects (2nd ed.). Cambridge, Massachusetts: Harvard Law School.

English, P. B., Olmedo, L., Bejarano, E., Lugo, H., Murillo, E., Seto, E., . . . Meltzer, D. (2017). The Imperial County Community Air Monitoring Network: a model for community-based environmental monitoring for public health action. *Environmental health perspectives, 125*(7), 074501.

EU-EOM. (2018). General Elections 2017 in Kenya, Final Report. Nairobi, Kenya: European Union Election Observation Mission.

European Commission. (2014). First Citizens' Observatories Projects Coordination Workshop. Retrieved from http://ec.europa.eu/research/environment/index_en.cfm?pg=gepw-meeting-5&section=geo

European Commission. (2015). Demonstrating the concept of 'Citizen Observatories'. Retrieved from https://cordis.europa.eu/programme/rcn/664594_en.html

European Commission. (2017). Europe's Digital Progress Report (EDPR) 2017, Country Profile The Netherlands. Brussels European Commision.

Fernandez-Gimenez, M., Ballard, H., & Sturtevant, V. (2008). Adaptive management and social learning in collaborative and community-based monitoring: a study of five community-based forestry organizations in the western USA. *Ecology and Society, 13*(2).

Ferster, C. J., & Coops, N. C. (2013). A review of earth observation using mobile personal communication devices. *Computers & Geosciences, 51,* 339-349. doi:http://dx.doi.org/10.1016/j.cageo.2012.09.009

Fisher, R. J. (1993). Social desirability bias and the validity of indirect questioning. *Journal of consumer research, 20*(2), 303-315.

Flyvbjerg, B. (1998). *Rationality and power: Democracy in practice*: University of Chicago press.

Frensley, T., Alycia, C., Marc, S., Rebecca, J., Steven, G., Michelle, P., . . . Joey, H. (2017). Bridging the Benefits of Online and Community Supported Citizen Science: A Case Study on Motivation and Retention with Conservation-Oriented Volunteers. *Citizen Science: Theory and Practice, 2*(1), 4. Retrieved from doi:10.5334/cstp.84

Friedman, A. J. (Ed.) (2008). *Framework for evaluating impacts of informal science education projects. Report from a National Science Foundation Workshop.* . Washington, DC: National Science Foundation.

Fritz, S., See, L., Carlson, T., Haklay, M., Oliver, J. L., Fraisl, D., . . . West, S. (2019). Citizen science and the United Nations Sustainable Development Goals. *Nature Sustainability, 2*(10), 922-930. doi:10.1038/s41893-019-0390-3

Fung, A. (2006). Varieties of participation in complex governance. *Public administration review, 66*(s1), 66-75. doi:http://dx.doi.org/10.1111/j.1540-6210.2006.00667.x

Gharesifard, M., & Wehn, U. (2016a). To share or not to share: Drivers and barriers for sharing data via online amateur weather networks. *Journal of Hydrology, 535,* 181-190. doi:http://dx.doi.org/10.1016/j.jhydrol.2016.01.036

Gharesifard, M., & Wehn, U. (2016b). What Drives Citizens to Engage in ICT-Enabled Citizen Science?: Case Study of Online Amateur Weather Networks, *Analyzing*

178

the Role of Citizen Science in Modern Research (pp. 62-88). Hershey, PA, USA: IGI Global.

Gharesifard, M., Wehn, U., & van der Zaag, P. (2017). Towards benchmarking citizen observatories: Features and functioning of online amateur weather networks. *Journal of Environmental Management, 193,* 381-393. doi:https://doi.org/10.1016/j.jenvman.2017.02.003

Gharesifard, M., Wehn, U., & van der Zaag, P. (2019a). Context matters: a baseline analysis of contextual realities for two community-based monitoring initiatives of water and environment in Europe and Africa. *Journal of Hydrology,* 124144. doi:https://doi.org/10.1016/j.jhydrol.2019.124144

Gharesifard, M., Wehn, U., & van der Zaag, P. (2019b). What influences the establishment and functioning of community-based monitoring initiatives of water and environment? A conceptual framework. *Journal of Hydrology, 579,* 124033. doi:https://doi.org/10.1016/j.jhydrol.2019.124033

Giesen, R. (2018). Ground Truth 2.0 project deliverable reports; Deliverable D2.9; Customized platform for the Netherlands Demo Case (Second Version). Delft, the Netherlands.

Gigler, B.-S., & Bailur, S. (2014). *Closing the feedback loop : can technology bridge the accountability gap?* : the World Bank Publications.

Gilardoni, L., Mazza, S., Ferraro, M., Carvalho, R., & Lobbrecht, A. (2013). Methodology for the e-collaboration environment, WeSenseIt project deliverable D.4.10. Sheffield, UK.

Gouveia, C., & Fonseca, A. (2008). New approaches to environmental monitoring: the use of ICT to explore volunteered geographic information. *GeoJournal, 72*(3-4), 185-197.

Government of Kenya. (1999). *The Environmental management and Coordination Act (No 8 of 1999).* Nairobi: The Government Printer.

Government of Kenya. (2010). *The Constitution of Kenya, 2010.* Nairobi: The Government Printer.

Government of Kenya. (2012). *County Governments Act (No 12 of 2012).* Nairobi: The Government Printer.

Government of Kenya (2013). *Agriculture and Food Authority Act (No 13 of 2013).* Nairobi: The Government Printer.

Government of Kenya. (2013). *Wildlife Conservation and Management (No 47 of 2013).* Nairobi: The Government Printer.

Government of Kenya. (2015). *Public Service (Values and Principles) Act (No 1A of 2015).* Nairobi: The Government Printer.

Government of Kenya. (2016). *Access To Information Act (No 31 of 2016).* Nairobi: The Government Printer.

Government of Kenya (2016). *The Public Participation Bill (Senate bills No. 15, 2016)*. Nairobi: The Government Printer.

Grandvoinnet, H., Aslam, G., & Raha, S. (2015). *Opening the black box: The contextual drivers of social accountability*: World Bank Publications.

Ground Truth 2.0 consortium. (2015). The Ground Truth 2.0 - Environmental knowledge discovery of human sensed data (Project proposal) *H2020-SC5-2015-two-stage, Topic SC5-17-2015, Innovation Actions*.

Guijt, I. (2014). Feedback loops–new buzzword, old practice? Retrieved from http://betterevaluation.org/blog/feedback_loops_new_buzzword_old_practice

Haklay, M. (2013). Citizen Science and Volunteered Geographic Information: Overview and Typology of Participation *Crowdsourcing geographic knowledge* (pp. 105-122). Netherlands: Springer

Haklay, M. (2014). citizen science definition. Retrieved from https://povesham.wordpress.com/category/citizen-science-2/

Haklay, M. (2015). Citizen Science and Policy: A European Perspective. Washington, USA: The Wodrow Wilson Center, Commons Lab.

Havekes, H., Koster, M., Dekking, W., Uijterlinde, R., Wensink, W., & Walkier, R. (2017). Water governance: The Dutch water authority model. *The Hague, NL, Dutch Water Authorities*.

Hecker, S., Wicke, N., Haklay, M., & Bonn, A. (2019). How Does Policy Conceptualise Citizen Science? A Qualitative Content Analysis of International Policy Documents. *Citizen Science: Theory and Practice, 4*(1).

Homewood, K. M. (2004). Policy, environment and development in African rangelands. *Environmental Science & Policy, 7*(3), 125-143.

Hsu, A., Malik, O., Johnson, L., & Esty, D. C. (2014). Development: Mobilize citizens to track sustainability. *Nature, 508*(7594), 33-35. doi:http://dx.doi.org/10.1038/508033a

Intelecon. (2016). ICT Access Gaps Study - Final Report. Nairobi, Kenya: Communications Authority of Kenya.

Irwin, A. (1995). *Citizen science : a study of people, expertise, and sustainable development*. London: Routledge.

Irwin, A. (2001). Constructing the scientific citizen: science and democracy in the biosciences. *Public Understanding of Science, 10*(1), 1-18.

Irwin, A. (2015). Citizen Science and Scientific Citizenship: Same Words, Different Meanings? *Science Communication Today*, 29-38.

ITU. (2017a). Measuring the Information Society Report 2017, Volume 1. Geneva, Switzerland: International Telecommunication Union.

ITU. (2017b). Measuring the Information Society Report 2017, Volume 2. Geneva, Switzerland: International Telecommunication Union.

Jafarkarimi, H., Sim, A., Saadatdoost, R., & Hee, J. M. (2014). The impact of ICT on reinforcing citizens' role in government decision making. *International Journal of Emerging Technology and Advanced Engineering, 4*(1), 642-646.

Jordan, R. C., Ballard, H. L., & Phillips, T. B. (2012). Key issues and new approaches for evaluating citizen-science learning outcomes. *Frontiers in Ecology and the Environment, 10*(6), 307-309.

Kaufmann, M., Doorn-Hoekveld, W. v., Gilissen, H.-K., & Van Rijswick, H. (2016). *Analysing and evaluating flood risk governance in the Netherlands: drowning in safety*: Utrecht: STARFLOOD Consortium.

Kemerink, J., Mbuvi, D., & Schwartz, K. (2013). Governance shifts in the water services sector: a case study of the Zambia water services sector *Water Services Management and Governance: Lessons for a Sustainable Future* (pp. 3-12). London: IWA Publishing.

Keough, H. L., & Blahna, D. J. (2006). Achieving integrative, collaborative ecosystem management. *Conservation Biology, 20*(5), 1373-1382.

Kersbergen, A., Gil-Roldán, E., & Costa, N. (2019). Ground Truth 2.0 project deliverable reports; Deliverable D3.3: Sustainable Business models for the Ground Truth 2.0 products/services. Delft, the Netherlands.

Kieslinger, B., Schäfer, T., Heigl, F., Dörler, D., Richter, A., & Bonn, A. (2017). The Challenge of Evaluation: An Open Framework for Evaluating Citizen Science Activities.

Kieslinger, B., Schäfer, T., Heigl, F., Dörler, D., Richter, A., & Bonn, A. (2018). *Evaluating Citizen Science: Towards an open framework. Book Chapter In: Citizen Science – Innovation in Open Science, Society and Policy*. London: UCL Press.

Kimura, A. H., & Kinchy, A. (2016). Citizen science: Probing the virtues and contexts of participatory research. *Engaging Science, Technology, and Society, 2*, 331-361.

Kruger, I. (2010). *The role of public participation in sustainable river basin development: Environmental planning in southern Spain.* UNESCO-IHE Institute for Water Education, Delft, the Netherlands. (WM.10.02)

Kullenberg, C., & Kasperowski, D. (2016). What Is Citizen Science?--A Scientometric Meta-Analysis. *PloS one, 11*(1), 1-16.

Kythreotis, A. P., Mantyka-Pringle, C., Mercer, T. G., Whitmarsh, L. E., Corner, A., Paavola, J., . . . Castree, N. (2019). Citizen social science for more integrative and effective climate action: A science-policy perspective. *Frontiers in Environmental Science, 7*, 10.

Lanfranchi, V., Wrigley, S., Ireson, N., Ciravegna, F., & Wehn, U. (2014). *Citizens' Observatories for Situation Awareness in Flooding.* Paper presented at the 11th International ISCRAM Conference (Information Systems for Crisis and Response Management), University Park, Pennsylvania, USA, May 2014.

Law of 31 October 1991. (1991). The Government Information (Public Access) Act. The Hague: House of Representatives of the Netherlands.

Leeuwis, C., Cieslik, K. J., Aarts, M. N. C., Dewulf, A. R. P. J., Ludwig, F., Werners, S. E., & Struik, P. C. (2018). Reflections on the potential of virtual citizen science platforms to address collective action challenges: Lessons and implications for future research. *NJAS - Wageningen Journal of Life Sciences, 86-87*, 146-157. doi:https://doi.org/10.1016/j.njas.2018.07.008

Liu, H.-Y., Kobernus, M., Broday, D., & Bartonova, A. (2014). A conceptual approach to a citizens' observatory–supporting community-based environmental governance. *Environmental Health, 13*(1), 107.

Lu, Y., Nakicenovic, N., Visbeck, M., & Stevance, A. S. (2015). Policy: Five priorities for the UN Sustainable Development Goals. *Nature, 520*(7548), 432-433. doi:http://dx.doi.org/10.1038/520432a

Macintosh, A. (2004). *Characterizing e-participation in policy-making.* Paper presented at the 37th Annual Hawaii International Conference on System Sciences, 5-8 Jan. 2004, Hawaii, U.S.A.

Macintosh, A., & Coleman, S. (2003). Promise and Problems of E-Democracy: Challenges of online citizen engagement. *Organisation for Economic Co-operation and Development.* doi:http://dx.doi.org/10.1787/9789264019492-en

MacKenzie, D., & Wajcman, J. (1999). Introductory essay: the social shaping of technology (2nd ed.). Buckingham, UK: Open University Press.

Mansell, R., & Wehn, U. (Eds.). (1998). *Knowledge societies: information technology for sustainable development.* New York: Oxford University Press; published for and on behalf of The United Nations.

Marine, J. (2015). *A Preliminary Comparative Study of Public Participation Acts in Kenya: A Case Study of Meru, Elgeyo/Marakwet & Machakos Counties.* Paper presented at the 38th AFSAAP Conference: 21st Century Tensions and Transformation, 28th-30th October, 2015, Melbourne, Australia.

Mayoux, L. (1995). Beyond Naivety: Women, Gender Inequality and Participatory Development. *DECH Development and Change, 26*(2), 235-258.

McCarthy, S., Tapsell, S., McDonagh, R., Lanfranchi, V., Anema, K., & Wehn de Montalvo, U. (2013). *Deliverable 6.21; Requirement analysis for citizen observatories (including stakeholder sensor adoption and usage).* Retrieved from

McElfish, J., Pendergrass, J., & Fox, T. (2016). Clearing the Path: Citizen Science and Public Decision Making in the United States. *Woodrow Wilson Center Science and Technology Innovation Program. Research Series, 4.*

McGuirk, P. M. (2001). Situating communicative planning theory: context, power, and knowledge. *Environment and Planning A, 33*(2), 195-217.

Michels, A. (2006). Citizen participation and democracy in the Netherlands. *Democratization, 13*(2), 323-339. doi:http://dx.doi.org/10.1080/13510340500524067

Ministry of Devolution and Planning & Council of Governors. (2016). *The County Public Participation Guidelines* Nairobi: The Government of Kenya.

Ministry of Economic Affairs. (2016). Digital agenda for the Netherlands innovation, trust, acceleration. The Hauge: European Commision.

Ministry of Justice and Security. (2010). *Safety Regions Act*. The Hague: Ministry of Justice and Security.

Ministry of Transport Public Works and Water Management. (2010). *Water Act*. The Hague, The Netherlands: Ministry of Transport, Public Works and Water Management.

Mokorosi, P. S., & van der Zaag, P. (2007). Can local people also gain from benefit sharing in water resources development? Experiences from dam development in the Orange-Senqu River Basin. *Physics and Chemistry of the Earth, Parts A/B/C, 32*(15–18), 1322-1329. doi:http://dx.doi.org/10.1016/j.pce.2007.07.028

Muli, D., & Mbelati, L. (2019). Summary of Conservancies. *Voice of the Mara, 5th Edition, July 2019*

Murphree, M. W. (2000). *Boundaries and borders: the question of scale in the theory and practice of common property management.* Paper presented at the Eighth Biennial Conference of the International Association for the Study of Common Property.

Nabatchi, T. (2012). *A manager's guide to evaluating citizen participation*: IBM Center for the Business of Government Washington, DC.

National Research Council. (2009). *Learning science in informal environments: People, places, and pursuits*. Washington, DC: The National Academies Press.

Nature. (2015). Rise of the citizen scientist. *Nature, 524*(7565), 265-265. doi:http://dx.doi.org/10.1038/524265a

Nelkin, D. (1975). The Political Impact of Technical Expertise. *Social Studies of Science, 5*(1), 35-54.

Nelson, N., & Wright, S. (1995). *Power and participatory development: theory and practice*: Intermediate Technology Publications Ltd (ITP).

Newman, G., Graham, J., Crall, A., & Laituri, M. (2011). The art and science of multi-scale citizen science support. *Ecological Informatics, 6*(3-4), 217-227.

Newman, G., Wiggins, A., Crall, A., Graham, E., Newman, S., & Crowston, K. (2012). The future of citizen science: emerging technologies and shifting paradigms. *Frontiers in Ecology and the Environment, 10*(6), 298-304. doi:10.1890/110294

Ngugi, E., Kipruto, S., & Samoei, P. (2013). Exploring Kenya's Inequality, Pulling Apart, or Pooling Together? *Narok County*. Nairobi, Kenya: the Kenya National Bureau of Statistics (KNBS) and the Society for International Development (SID).

Novoa, S., & Wernand, M. (2013). Research methodology for user validations, Citclops project deliverable D.3.1 Barcelona, Spain.

183

OECD. (2012). *OECD Environmental Outlook to 2050: The Consequences of Inaction.* Retrieved from http://www.oecd.org/env/indicators-modelling-outlooks/oecdenvironmentaloutlookto2050theconsequencesofinaction.htm

OECD. (2014). Water Governance in the Netherlands: Fit for the Future? : OECD Studies on Water, OECD Publishing. http://dx.doi.org/10.1787/9789264102637-en.

OMNISCIENTIS Consortium. (2014). Final Report Summary, OMNISCIENTIS project deliverable. Angleur, Belgium.

Omoto, L., Gharesifard, M., & van der Kwast, H. (2018). Ground Truth 2.0 project deliverable reports; Deliverable D2.6; Customized platform for Kenyan Demo Case (First Version). Delft, the Netherlands.

Onencan, A. M., Meesters, K., & Van de Walle, B. (2018). Methodology for participatory gis risk mapping and citizen science for solotvyno salt mines. *Remote Sensing, 10*(11), 1828.

OpenScientist blog. (2011). Finalizing a Definition of "Citizen Science" and "Citizen Scientists". Retrieved from http://www.openscientist.org/2011/09/finalizing-definition-of-citizen.html

Osterwalder, A., & Pigneur, Y. (2010). *Business model generation: a handbook for visionaries, game changers, and challengers.* Hoboken, New Jersey: John Wiley & Sons.

Oxford English Dictionary. (2014). citizen science definition. Retrieved from http://www.oxforddictionaries.com/definition/english/citizen-science

Pahl-Wostl, C. (2009). A conceptual framework for analysing adaptive capacity and multi-level learning processes in resource governance regimes. *Global Environmental Change, 19*(3), 354-365. doi:http://dx.doi.org/10.1016/j.gloenvcha.2009.06.001

Phillips, T., Bonney, R., & Shirk, J. (2012). What is our impact? Toward a Unified Framework for Evaluating Outcomes of Citizen Science Participation. *Citizen science: Public participation in environmental research*, 82-95.

Phillips, T., Ferguson, M., Minarchek, M., Porticella, N., & Bonney, R. (2014). Evaluating learning outcomes from citizen science. Ithaca, New York: Cornell Lab of Ornithology.

Phillips, T., Porticella, N., Constas, M., & Bonney, R. (2018). A Framework for Articulating and Measuring Individual Learning Outcomes from Participation in Citizen Science. *Citizen Science: Theory and Practice, 3*(2).

Pimm, S. L., Alibhai, S., Bergl, R., Dehgan, A., Giri, C., Jewell, Z., . . . Loarie, S. (2015). Emerging technologies to conserve biodiversity. *Trends in Ecology & Evolution, 30*(11), 685-696.

Pocock, M. J., Chapman, D. S., Sheppard, L. J., & Roy, H. E. (2014). A strategic framework to support the implementation of citizen science for environmental monitoring. Final report to SEPA.

Pollock, R. M., & Whitelaw, G. S. (2005). Community-based monitoring in support of local sustainability. *Local Environment, 10*(3), 211-228.

Reed, M. S. (2008). Stakeholder participation for environmental management: A literature review. *Biological Conservation, 141*(10), 2417-2431. doi:http://dx.doi.org/10.1016/j.biocon.2008.07.014

Richter, A., Hauck, J., Feldmann, R., Kühn, E., Harpke, A., Hirneisen, N., . . . Bonn, A. (2018). The social fabric of citizen science—drivers for long-term engagement in the German butterfly monitoring scheme. *Journal of insect conservation, 22*(5-6), 731-743.

Roy, H., Pocock, M., Preston, C., Roy, D., Savage, J., Tweddle, J., & Robinson, L. (2012). *Understanding citizen science and environmental monitoring: Final Report on behalf of UK-EOF*. NERC Centre for Ecology & Hydrology and Natural History Museum.

Rubio Iglesias, J. M. (2015). *Citizens' Observatories & Crowdsourcing, novel ways to engage citizens in science and environmental policy-making*. Paper presented at the Geospatial World Forum-INSPIRE Conference, Lisbon.

Rutten, M., Minkman, E., & van der Sanden, M. (2017). How to get and keep citizens involved in mobile crowd sensing for water management? A review of key success factors and motivational aspects. *Wiley Interdisciplinary Reviews: Water, 4*(4), e1218.

Sachs, J. D., Modi, V., Figueroa, H., Fantacchiotti, M. M., Sanyal, K., Khatun, F., & Shah, A. (2015). ICT & SDGs - How Information and Communications Technology Can Achieve The Sustainable Development Goals: Earth Institute at Columbia University in collaboration with Ericsson.

Sauti za Wananchi. (2018). Active and engaged? Kenyans' views and experiences on citizen participation, *Research Brief No. 24*. Nairobi, Kenya: Twaweza East Africa.

Schäfer, T., & Kieslinger, B. (2016). Supporting emerging forms of citizen science: A plea for diversity, creativity and social innovation. *Journal of Science Communication, 15*(02), Y02.

Shirk, J., Ballard, H., Wilderman, C., Phillips, T., Wiggins, A., Jordan, R., . . . Krasny, M. (2012). Public participation in scientific research: a framework for deliberate design. *Ecology and Society, 17*(2).

Silvertown, J. (2009). A new dawn for citizen science. *Trends in Ecology & Evolution, 24*(9), 467-471. doi:http://dx.doi.org/10.1016/j.tree.2009.03.017

Tredick, C. A., Lewison, R. L., Deutschman, D. H., Hunt, T. A., Gordon, K. L., & Von Hendy, P. (2017). A rubric to evaluate citizen-science programs for long-term ecological monitoring. *BioScience, 67*(9), 834-844.

Tulloch, A. I., Auerbach, N., Avery-Gomm, S., Bayraktarov, E., Butt, N., Dickman, C. R., . . . Holden, M. H. (2018). A decision tree for assessing the risks and benefits of publishing biodiversity data. *Nature ecology & evolution, 2*(8), 1209-1217.

UNDP. (1992). *Rio Declaration on Environment and Development.* United Nations Development Programme Retrieved from http://www.unep.org/Documents.Multilingual/Default.asp?documentid=78&articleid=1163.

UNECE. (1998). *Convention on Access to Information, Public Participation in decision-making and access to justice in environmental matters.* Aarhus, Denmark: The United Nations Economic Commission for Europe (UNECE) Retrieved from http://www.unece.org/fileadmin/DAM/env/pp/documents/cep43e.pdf.

UNISDR. (2005). *Hyogo Framework for Action 2005–2015: Building the Resilience of Nations and Communities to Disasters.* United Nations International Strategy for Disaster Reduction Retrieved from http://www.unisdr.org/we/coordinate/hfa.

United Nations. (2012). *Renewable Resources and Conflict, .* Retrieved from http://www.un.org/en/land-natural-resources-conflict/renewable-resources.shtml

United Nations. (2015). *The Sustainable Development Goals (SDGs).* Division for Sustainable Development, Department of Economic and Social Affairs, United Nations Retrieved from https://sustainabledevelopment.un.org/sdgs.

United Nations. (2019). *World Population Prospects 2019: Highlights (ST/ESA/SER.A/423).* New York: Department for Economic and Social Affairs, Population Division.

van Deursen, A. J. A. M., & van Dijk, J. A. G. M. (2015). Toward a Multifaceted Model of Internet Access for Understanding Digital Divides: An Empirical Investigation. *The Information Society, 31*(5), 379-391. doi:https://doi.org/10.1080/01972243.2015.1069770

van Dijk, J. A. G. M. (2006). Digital divide research, achievements and shortcomings. *Poetics, 34*(4–5), 221-235. doi:http://dx.doi.org/10.1016/j.poetic.2006.05.004

Van Wee, B., & Banister, D. (2016). How to write a literature review paper? *Transport Reviews, 36*(2), 278-288.

Vergouwe, R. (2016). *The national flood risk analysis for the netherlands*: Rijkswaterstaat VNK Project Office.

Videira, N., Antunes, P., Santos, R., & Lobo, G. (2006). Public and stakeholder participation in European water policy: a critical review of project evaluation processes. *European Environment, 16*(1), 19-31.

Warburton, D., Dudding, J., Sommer, F., & Walker, P. (2001). Evaluating participatory, deliberative and co-operative ways of working. *Brighton: Crown.*

Warner, J. F. (2006). More sustainable participation? Multi-stakeholder platforms for integrated catchment management. *Water resources development, 22*(1), 15-35.

Wehn de Montalvo, U., Evers, J., Rusca, M., Faedo, G., & Onencan, A. (2013). Report on the governance context for the citizen observatories of water, WeSenseIt project deliverable D6.10. Delft, the Netherlands.

Wehn, U., Alfonso, L., Lobbrecht, A., van de Giesen, N., van der Kwast, H., Joshi, S., & Masó, J. (2015a). Ground Truth 2.0 - Environmental knowledge discovery of human sensed data, Horizon 2020 project, Description of Action.

Wehn, U., & Almomani, A. (2019). Incentives and barriers for participation in community-based environmental monitoring and information systems: A critical analysis and integration of the literature. *Environmental Science & Policy, 101*, 341-357. doi:https://doi.org/10.1016/j.envsci.2019.09.002

Wehn, U., Anema, K., & Gharesifard, M. (2016). Report on social innovation and impact of citizen observatory-based knowledge exchange and participation, WeSenseIt project deliverable D6.3. Delft, the Netherlands.

Wehn, U., & Evers, J. (2014). *Citizen observatories of water: Social innovation via eParticipation?* Paper presented at the ICT for Sustainability 2014 (ICT4S-14).

Wehn, U., Pfeiffer, E., Gharesifard, M., Alfonso, L., & Anema, K. (2020). Updated validation and socio-economic impacts report, Ground Truth 2.0 project deliverable D1.12. Delft, the Netherlands.

Wehn, U., Pfeiffer, E., Gharesifard, M., Anema, K., & Remmers, M. (2017). Methodology for validation and impact assessment, Ground Truth 2.0 project deliverable D1.10. Delft, the Netherlands.

Wehn, U., Rusca, M., Evers, J., & Lanfranchi, V. (2015b). Participation in flood risk management and the potential of citizen observatories: A governance analysis. *Environmental Science & Policy, 48*(0), 225-236. doi:http://dx.doi.org/10.1016/j.envsci.2014.12.017

WeSenseIt Consortium. (2016). Citizen observatory policy brief - citizen participation in the digital age – from policy to practice, WeSenseIt project deliverable D8.43. Delft, the Netherlands.

Whitelaw, G., Vaughan, H., Craig, B., & Atkinson, D. (2003). Establishing the Canadian Community Monitoring Network. *Environmental Monitoring and Assessment, 88*(1), 409-418. doi:10.1023/a:1025545813057

Wiggins, A., Bonney, R., LeBuhn, G., Parrish, J. K., & Weltzin, J. F. (2018). A Science Products Inventory for Citizen-Science Planning and Evaluation. *BioScience, 68*(6), 436-444. doi:10.1093/bioscience/biy028

Wiggins, A., & Crowston, K. (2011). *From conservation to crowdsourcing: A typology of citizen science.* Paper presented at the 44th hawaii international conference on system sciences, 04 - 07 January 2011, Kauai, HI, USA.

Wiggins, A., & Crowston, K. (2012). *Goals and tasks: Two typologies of citizen science projects.* Paper presented at the 44th Hawaii International Conference on System Sciences (HICSS), 04 - 07 January, 2011, Kauai, HI, USA.

Winner, L. (1980). Do Artifacts Have Politics? *Daedalus, 109*(1), 121-136.

Winner, L. (1986). *The whale and the reactor : a search of limits in an age of high technology.* Chicago (Ill.); London: The University of Chicago Press.

World Bank Group. (2014). Strategic Framework for Mainstreaming Citizen Engagement in World Bank Group Operations. Washington, DC. © World Bank. https://openknowledge.worldbank.org/handle/10986/21113.

World Bank Group. (2016). Evaluating Digital Citizen Engagement. Washington, DC. © World Bank. https://openknowledge.worldbank.org/handle/10986/23752.

World Economic Forum. (2016a). The Global Information Technology Report 2016 *Innovating in the Digital Economy*. Geneva, Switzerland: World Economic Forum.

World Economic Forum. (2016b). The Global Risks Report 2016, 11th Edition. Geneva: World Economic Forum, The Global Competitiveness and Risks Team.

World Economic Forum. (2019). The Global Risks Report 2019, 14th Edition. Geneva: World Economic Forum, The Global Competitiveness and Risks Team.

World Wide Web Foundation. (2015). Women's Rights Online, Translating Access into Empowerment. Geneva: World Wide Web Foundation

Yamada, F., Kakimoto, R., Yamamoto, M., Fujimi, T., & Tanaka, N. (2011). Implementation of community flood risk communication in Kumamoto, Japan. *Journal of advanced transportation, 45*(2), 117-128. doi:http://dx.doi.org/10.1002/atr.119

Yin, R. K. (1984). *Case study research : design and methods*. Beverly Hills, Calif.: Sage Publications.

ANNEX 1: OBSERVATION PROTOCOL

Code:	Reference audio/video record:
Date:	Start time: Finish time:

Title of event/observation opportunity:

Goals

Physical surroundings

Characteristics of participants (individually and as a group)

Facilitation

Interactions (collective)

Nonverbal behavior
Direct quote(s)
Other observations

ANNEX 2: LIST OF PHASE 1 INTERVIEW QUESTIONS IN THE NETHERLANDS CASE STUDY

The interview protocol included two parts; Part I focused on the issue in focus of the CBM initiative and Part II included generic demographic questions. A slightly different set of context-related questions (question in Part I) was used to interview each stakeholder category. The box underneath each question clarifies which stakeholders were asked to answer each question in Part I. All respondents answered the questions in Part II. Because of the complexity of the topic, for some questions we needed to prompt the interviewee to provide relevant answers, therefore where relevant, these prompts are presented after the questions.

Part I - context-related questions

1. **In the Netherlands, how urgent do you think is the need for reducing/ preventing local flooding?**

   Participants in the co-design meetings ☒   Regulatory entities ☒   General public ☒   Expert advisors ☒

2. **In terms of percentage, how many Dutch citizens would agree with you on that level of urgency for reducing/ preventing local flooding?**

   Participants in the co-design meetings ☒   Regulatory entities ☒   General public ☒   Expert advisors ☒

3. **Which stakeholders are involved in policy making regarding local flood management in the Netherlands?**

   [Prompt] Mention involved stakeholders at different levels (International/national/regional/local)

   Participants in the co-design meetings ☐   Regulatory entities ☒   General public ☐   Expert advisors ☒

191

4. **What are the formal institutions and policies related to local flood management in the Netherlands? To what extent are they being implemented?**

    [Prompt] Formal institutions are the rules and strategies that govern the decision making processes

Participants in the co-design meetings ☐ Regulatory entities ☒ General public ☐ Expert advisors ☒

5. **Which stakeholders are involved in decision making processes regarding local flood management in the Netherlands?**

    [Prompt] Mention involved stakeholders at different levels (International/national/regional/local)

Participants in the co-design meetings ☐ Regulatory entities ☒ General public ☐ Expert advisors ☒

6. **Regarding local flood management in the Netherlands: what is the hierarchy of authority between supranational, national, provincial, and local entities?**

    [Prompt] Mention how these entities interact with one another at different levels

Participants in the co-design meetings ☐ Regulatory entities ☒ General public ☐ Expert advisors ☒

7. **In what ways, if any, you are involved in reducing/preventing local floods in your place of residence?**

    [Prompt] What specific aspect(s)? At what level (international, national, local, etc.)?

Participants in the co-design meetings ☒ Regulatory entities ☒ General public ☒ Expert advisors ☒

8. **How do you take part in decision making about reducing/preventing local floods in your place of residence?**

    [Prompt] Taking part can have many forms, e.g. negotiating, deliberating or bargaining with different stakeholders, or having a direct say in the decision making or implementation)

Participants in the co-design meetings ☒ Regulatory entities ☒ General public ☒ Expert advisors ☒

9. **What influence, if any, do you think you have on these decisions?**

    [Prompt] Influence can be e.g. via having an official mandate, joining officials for making decisions, providing advice or consult to decision makers, or influencing public opinion

    Participants in the co-design meetings ☒   Regulatory entities ☒   General public ☒   Expert advisors ☒

10. **What kind of efforts would you need to put in if you would want to participate in decision making processes about reducing/preventing local floods in your place of residence?**

    [Prompt]: e.g. Time, Money, tools, skills, etc.

    Participants in the co-design meetings ☒   Regulatory entities ☒   General public ☒   Expert advisors ☒

11. **Do you know of any duplication of efforts or overlapping rules/roles/responsibilities about flood management in your place of residence, or in the Netherlands? If yes, what are they?**

    Participants in the co-design meetings ☒   Regulatory entities ☒   General public ☒   Expert advisors ☒

12. **What values, norms, traditions come into play when managing local floods in the Netherlands?**

    Participants in the co-design meetings ☒   Regulatory entities ☒   General public ☒   Expert advisors ☒

13. **How strictly implemented are the rules, roles and responsibilities regarding local flood management in the Netherlands?**

    Participants in the co-design meetings ☒   Regulatory entities ☒   General public ☒   Expert advisors ☒

14. What communication channels do you most use for sharing information with others? How often do you use them?

[Prompt]: e.g. Face-to-Face, telephone (call or SMS), email, Websites or Blogs, Social media (Facebook, Twitter, etc.), Apps on my Smartphone (WhatsApp, Viber, Line, etc.), Mass and print media (radio, TV, etc.)

Participants in the co-design meetings ☒   Regulatory entities ☒   General public ☒   Expert advisors ☒

15. Do you communicate about local flooding with others? With whom? How?

[Prompt] By communicate we mean exchange of information that can have many forms, e.g. listening to discussions, discussing with others, expressing preferences, or sharing data and information

Participants in the co-design meetings ☒   Regulatory entities ☒   General public ☒   Expert advisors ☒

16. What channel(s) do you prefer to use for communicating about local flooding (e.g. with other citizens, scientists, decision makers, etc.)?

[Prompt] Face-to-Face, telephone (call or SMS), email, websites or Blog, social media (e.g. Facebook), using apps on your Smartphone (e.g. WhatsApp), or any other way

- **Face-to-face**
- **Via Telephone (call or SMS)**
- **Via email**
- **Using a website or blog**
- **Via social media**
- **Using an App on my Smartphone**
- **Other**

Participants in the co-design meetings ☒   Regulatory entities ☒   General public ☒   Expert advisors ☒

17. How do you assess the availability of local flooding data in the Netherlands in terms of location and time? What do you use it for?

Participants in the co-design meetings ☒   Regulatory entities ☒   General public ☒   Expert advisors ☒

18. Who do you think has the required skills and experience to analyse local flood data in the Netherlands?

[Prompt] e.g. organizations or individuals

Participants in the co-design meetings ☒   Regulatory entities ☒   General public ☒   Expert advisors ☒

19. Who defines the level of access to local flooding data for different stakeholders in the Netherlands?

Participants in the co-design meetings ☐   Regulatory entities ☒   General public ☐   Expert advisors ☒

20. What personal experience, if any, do you have with accessing data and information about local floods in the Netherlands?

[Prompt]: e.g. availability, accessibility, reliability, quality, etc. of the data/information

Participants in the co-design meetings ☒   Regulatory entities ☒   General public ☒   Expert advisors ☒

21. Do you have any other comments, questions, or concerns that you would like to share with us?

Participants in the co-design meetings ☒   Regulatory entities ☒   General public ☒   Expert advisors ☒

Part II - Demographic questions

22. What is your age?
- Under 18 years
- 18 to 25 years
- 26 to 35 years
- 36 to 45 years
- 46 to 55 years
- 56 to 65 years
- 66 years or older
- I prefer not to answer

23. **What is your gender?**
    - Male
    - Female
    - I prefer not to answer

24. **What is the highest degree or level of education you have completed?**
    - Less than high school
    - High school graduate
    - Completed some college
    - Associate degree
    - Bachelor's degree
    - Completed some postgraduate courses
    - Master's degree
    - Ph.D.
    - Prefer not to answer

25. **What best describes your current work situation?**
    - I work 30 hours or more per week
    - I work less than 30 hours per week
    - I am not currently employed
    - Prefer not to answer

26. **What is your present occupation?**
27. **Where do you currently live?**
28. **What is your email address?**

ANNEX 3: LIST OF PHASE 1 INTERVIEW QUESTIONS IN THE KENYA CASE STUDY

The interview protocol included two parts; Part I focused on the issue in focus of the CBM initiative and Part II included generic demographic questions. A slightly different set of context-related questions (question in Part I) was used to interview each stakeholder category. The box underneath each question clarifies which stakeholders were asked to answer each question in Part I. All respondents answered the questions in Part II. Because of the complexity of the topic, for some questions we needed to prompt the interviewee to provide relevant answers, therefore where relevant, these prompts are presented after the questions.

Part I - context-related questions

1. **In Kenya, how urgent do you think is the need for sustainable biodiversity management? What about improving people's livelihood?**

Participants in the co-design meetings ☒   Regulatory entities ☒   General public ☒   Expert advisors ☒

2. **In terms of percentage, how many Kenyans would agree with you on that level of urgency for biodiversity management? What about improving people's livelihood?**

Participants in the co-design meetings ☒   Regulatory entities ☒   General public ☒   Expert advisors ☒

3. **Which stakeholders are involved in policy making regarding biodiversity management in Kenya? What about livelihoods?**

[Prompt] Mention involved stakeholders at different levels (International/national/regional/local)

Participants in the co-design meetings ☐   Regulatory entities ☒   General public ☐   Expert advisors ☒

197

4. What are the formal institutions and policies related to biodiversity management and sustainable livelihoods in Kenya? To what extent are they being implemented?

[Prompt] Formal institutions are the rules and strategies that govern the decision making processes

Participants in the co-design meetings ☐   Regulatory entities ☒   General public ☐   Expert advisors ☒

5. Which stakeholders are involved in decision making processes regarding biodiversity management and sustainable livelihoods in Kenya?

[Prompt] Mention involved stakeholders at different levels (International/national/regional/local)

Participants in the co-design meetings ☐   Regulatory entities ☒   General public ☐   Expert advisors ☒

6. Regarding biodiversity and sustainable livelihood management in Kenya: what is the hierarchy of authority between supranational, national, provincial, and local entities?

[Prompt] Mention how these entities interact with one another at different levels

Participants in the co-design meetings ☐   Regulatory entities ☒   General public ☐   Expert advisors ☒

7. In what ways, if any, you are involved in biodiversity and/or sustainable livelihoods management in the Mara region?

[Prompt] What specific aspect(s)? At what level (international, national, local, etc.)?

Participants in the co-design meetings ☒   Regulatory entities ☒   General public ☒   Expert advisors ☒

8. How do you take part in decision making about biodiversity management and/or sustainable livelihoods in Kenya?

[Prompt] Taking part can have many forms, e.g. negotiating, deliberating or bargaining with different stakeholders, or having a direct say in the decision making or implementation)

Participants in the co-design meetings ☒   Regulatory entities ☒   General public ☒   Expert advisors ☒

198

9. **What influence, if any, do you think you have on these decisions?**

   [Prompt] Influence can be e.g. via having an official mandate, joining officials for making decisions, providing advice or consult to decision makers, or influencing public opinion

   Participants in the co-design meetings ☒   Regulatory entities ☒   General public ☒   Expert advisors ☒

10. **What kind of efforts would you need to put in if you would want to participate in decision making processes about biodiversity management in Kenya? What about decisions on livelihoods?**

    [Prompt]: e.g. Time, Money, tools, skills, etc.

    Participants in the co-design meetings ☒   Regulatory entities ☒   General public ☒   Expert advisors ☒

11. **Do you know of any duplication of efforts or overlapping rules/roles/responsibilities about biodiversity management or sustainable livelihoods in Kenya? If yes, what are they?**

    Participants in the co-design meetings ☒   Regulatory entities ☒   General public ☒   Expert advisors ☒

12. **What values, norms, traditions come into play when managing biodiversity in Kenya? What about livelihoods?**

    Participants in the co-design meetings ☒   Regulatory entities ☒   General public ☒   Expert advisors ☒

13. **How strictly implemented are the rules, roles and responsibilities regarding biodiversity management in Kenya? What about people's livelihoods?**

    Participants in the co-design meetings ☒   Regulatory entities ☒   General public ☒   Expert advisors ☒

14. What communication channels do you most use for sharing information with others? How often do you use them?

[Prompt]: e.g. Face-to-Face, telephone (call or SMS), email, Websites or Blogs, Social media (Facebook, Twitter, etc.), Apps on my Smartphone (WhatsApp, Viber, Line, etc.), Mass and print media (radio, TV, etc.)

Participants in the co-design meetings ☒   Regulatory entities ☒   General public ☒   Expert advisors ☒

15. Do you communicate about biodiversity management with others? With whom? How? What about communicating about livelihoods?

[Prompt] By communicate we mean exchange of information that can have many forms, e.g. listening to discussions, discussing with others, expressing preferences, or sharing data and information

Participants in the co-design meetings ☒   Regulatory entities ☒   General public ☒   Expert advisors ☒

16. What channel(s) do you prefer to use for communicating about biodiversity with others (e.g. other citizens, scientists, decision makers, etc.)? What about data on livelihoods?

[Prompt] Face-to-Face, telephone (call or SMS), email, websites or Blog, social media (e.g. Facebook), using apps on your Smartphone (e.g. WhatsApp), or any other way

- **Face-to-face**
- **Via Telephone (call or SMS)**
- **Via email**
- **Using a website or blog**
- **Via social media**
- **Using an App on my Smartphone**
- **Other**

Participants in the co-design meetings ☒   Regulatory entities ☒   General public ☒   Expert advisors ☒

17. How do you assess the availability of biodiversity data in Kenya in terms of location and time? What do you use it for? What about data on livelihoods?

Participants in the co-design meetings ☒   Regulatory entities ☒   General public ☒   Expert advisors ☒

18. Who do you think has the required skills and experience to analyse biodiversity data in Kenya? What about data on livelihoods?

[Prompt] e.g. organizations or individuals

Participants in the co-design meetings ☒ Regulatory entities ☒ General public ☒ Expert advisors ☒

19. Who defines the level of access to biodiversity data for different stakeholders in Kenya? What about data on livelihoods?

Participants in the co-design meetings ☐ Regulatory entities ☒ General public ☐ Expert advisors ☒

20. What personal experience, if any, do you have with accessing data and information about biodiversity or livelihoods in Kenya?

[Prompt]: e.g. availability, accessibility, reliability, quality, etc. of the data/information

Participants in the co-design meetings ☒ Regulatory entities ☒ General public ☒ Expert advisors ☒

21. Do you have any other comments, questions, or concerns that you would like to share with us?

Participants in the co-design meetings ☒ Regulatory entities ☒ General public ☒ Expert advisors ☒

Part II - Demographic questions

22. What is your age?
- Under 18 years
- 18 to 25 years
- 26 to 35 years
- 36 to 45 years
- 46 to 55 years
- 56 to 65 years
- 66 years or older
- I prefer not to answer

23. What is your gender?
- Male
- Female
- I prefer not to answer

24. What is the highest degree or level of education you have completed?

- Less than high school
- High school graduate
- Completed some college
- Associate degree
- Bachelor's degree
- Completed some postgraduate courses
- Master's degree
- Ph.D.
- Prefer not to answer

25. What best describes your current work situation?

- I work 30 hours or more per week
- I work less than 30 hours per week
- I am not currently employed
- Prefer not to answer

26. What is your present occupation?

27. Where do you currently live?

What is your email address?

ANNEX 4: LIST OF PHASE 2 INTERVIEW QUESTIONS IN THE NETHERLANDS CASE STUDY

This interview protocol has six parts; Parts I to V are in line with the five dimensions of the CPI Framework and Part VI includes generic demographic questions. A slightly different set of questions (question in Part I to V) was used to interview each interviewee category. The box underneath each question clarifies which interviewees were asked to answer each question. All respondents answered the generic demographic questions in Part VI. Because of the complexity of the topic, for some questions we needed to prompt the interviewee to provide relevant answers, therefore where relevant, these prompts are presented after the questions.

Part I - Goals & Objectives

1. **In your view, what are the most important objectives of Grip op Water Altena?**

    CBM members ⊠ GT2.0 team members ⊠

2. **To what extent do you think the objectives of Grip op Water Altena are achieved so far?**

    CBM members ⊠ GT2.0 team members ⊠

3. **If you think there were shortcomings in achieving the objectives, what do you think were the reasons?**

    CBM members ⊠ GT2.0 team members ⊠

4. **How did you participate in Grip op Water Altena?**

    CBM members ⊠ GT2.0 team members ⊠

203

5. **Why did you participate in Grip op Water Altena?**

   [instructions] This question is only relevant if the DC lead or team members have participated in the CO beyond the 'regular' project responsibilities (e.g. participation in data collection and sharing).

   CBM members ☒   GT2.0 team members ☒

6. **How did you monitor the achievement of the objectives of Grip op Water Altena and the extent of their achievement?**

   CBM members ☐   GT2.0 team members ☒

7. **Were there any changes in the objectives of Grip op Water Altena? If yes, what were the changes and why did they happen?**

   [Prompt] Please think of both explicit and implicit (formal/informal) changes in the objectives.

   CBM members ☐   GT2.0 team members ☒

Part II - Technology

8. **What is the 'current' architecture of technologies used in Grip op Water Altena?**

   CBM members ☐   GT2.0 team members ☒

9. **What were the priority of functionalities and matching tools for Grip op Water Altena?**

   CBM members ☐   GT2.0 team members ☒

10. **If there are key differences between the initial and current set of technologies, why did this happen?**

    CBM members ☐   GT2.0 team members ☒

11. **Who do you think are the main target users of the tools developed in Grip op Water Altena?**

    CBM members ☒ GT2.0 team members ☒

12. **How accessible do you think the tools developed in Grip op Water Altena (e.g. App, website, sensors, test kits, etc.] are for the target users?**

    [Prompt] You can think of accessibility in terms of physical access to technologies or internet or e.g. required skills to use the tools.

    CBM members ☒ GT2.0 team members ☒

13. **Which stakeholder groups were involved in designing the functionalities of Grip op Water Altena?**

    CBM members ☐ GT2.0 team members ☒

14. **Which stakeholder groups were not involved in designing the functionalities of Grip op Water Altena?**

    CBM members ☐ GT2.0 team members ☒

15. **Who are the current end-users of the tools (Apps, web-platforms, etc.) developed in Grip op Water Altena (demographic info, stakeholder category, etc.)?**

    CBM members ☐ GT2.0 team members ☒

16. **Which stakeholder groups are currently not using the tools (Apps, web-platforms, etc.) developed in Grip op Water Altena (demographic info, stakeholder category, etc.)?**

    CBM members ☐ GT2.0 team members ☒

17. **Do you have any difficulty accessing or using the tools developed in Grip op Water Altena (e.g. App, website, sensors, test kits, etc.)? If yes, which one(s) and why?**

CBM members   ☒   GT2.0 team members   ☐

<u>Part III - Power dynamics</u>

18. **Since the start of this initiative in May 2017 have you noticed any changes in the social, institutional & political context, which are relevant for reducing/preventing local flooding in Land van Heusden en Altena? If yes, what were the changes?**

CBM members   ☒   GT2.0 team members   ☒

19. **To what extent, if any, do you think your influence in decision making processes regarding reducing/preventing local flooding in Land van Heusden en Altena has changed because of your participation in Grip op Water Altena? Please explain why?**

[Prompt] Influence can be e.g. via having an official mandate, joining officials for making decisions, providing advice or consult to decision makers, or influencing public opinion.

CBM members   ☒   GT2.0 team members   ☒

20. **Do you use the data/information produced in Grip op Water Altena? If yes, what do you use this data/information for?**

CBM members   ☒   GT2.0 team members   ☒

21. To what extent do you think your access to data/information regarding local flooding in Land van Heusden en Altena has changed because of your participation in Grip op Water Altena?

CBM members ☒   GT2.0 team members ☒

22. T To what extent do you think you have control over the data/information that is being produced in Grip op Water Altena?

[Prompt] You can think of e.g. access to raw data, or the ability to influence data collection, quality control or data sharing processes.

CBM members ☒   GT2.0 team members ☒

23. How was Grip op Water Altena established?
- **Top-Down - Grip op Water Altena was initiated and controlled by authorities or stakeholders at higher levels of policy or decision making**
- **Bottom-Up - Grip op Water Altena was initiated and controlled by stakeholders such as citizens or volunteers who have no official mandate for policy or decision making**
- **Co-created - Grip op Water Altena was initiated and controlled by a group of interested stakeholders who had the chance to influence its design and functionalities**
- **Commerce Driven - Grip op Water Altena is a market-based and for-profit initiative**
- **Other**

CBM members ☒   GT2.0 team members ☒

24. What are the envisioned revenue stream for Grip op Water Altena??

CBM members ☒   GT2.0 team members ☒

207

Part IV - Participation

25. What kind of efforts do you need to put in to participate in Grip op Water Altena?

CBM members ☒   GT2.0 team members ☒

26. What kind of support is provided for participation in Grip op Water Altena?

CBM members ☒   GT2.0 team members ☒

27. What are the communication channels in Grip op Water Altena?

[Prompt] Face-to-face, telephone (call or SMS), email, build-in functionalities on the website, a blog, social media (e.g. Facebook), using apps on your smartphone (e.g. WhatsApp), or any other way.

CBM members ☐   GT2.0 team members ☒

28. To what extent you have so far used the following communication channels in Grip op Water Altena?

	Very frequently	Occasionally	Rarely	Never
Face-to-face				
Telephone (call)				
Telephone (SMS)				
Email				
Website or Blog				
Social Media (e.g. Facebook, Twitter, etc.)				
Apps on smartphone (e.g. WhatsApp)				
Mass and print media (e.g. radio, TV, etc.)				
Other				

CBM members ☒   GT2.0 team members ☐

Part V - Results

29. What did you expect to see as the outputs (direct products) of Grip op Water Altena?

[prompt] you can think of both positive and negative individual, societal, scientific, economic, environmental and governance-related outputs

CBM members ☒   GT2.0 team members ☒

30. What do you think are the actual outputs (direct products) of Grip op Water Altena?

[prompt] you can think of both positive and negative individual, societal, scientific, economic, environmental and governance-related outputs

CBM members ☒   GT2.0 team members ☒

31. What do you think are the actual short-term or incidental changes (outcomes) that have happened because of Grip op Water Altena?

[prompt] you can think of both positive and negative individual, societal, scientific, economic, environmental and governance-related short-term changes

CBM members ☒   GT2.0 team members ☒

32. What short-term or incidental changes (outcomes) do you expect to see in the near future as a result of Grip op Water Altena?

[prompt] you can think of both positive and negative individual, societal, scientific, economic, environmental and governance-related short-term changes

CBM members ☒   GT2.0 team members ☒

i.

33. What long-term changes (impacts) do you expect to see as a result of Grip op Water Altena?

[prompt] you can think of both positive and negative individual, societal, scientific, economic, environmental and governance-related long-term changes

CBM members ☒   GT2.0 team members ☒

Part VI - Demographic questions

34. What is your age?

- Under 18 years
- 18 to 25 years
- 26 to 35 years
- 36 to 45 years
- 46 to 55 years
- 56 to 65 years
- 66 years or older
- I prefer not to answer

35. What is your gender?

- Male
- Female
- I prefer not to answer

36. What is the highest degree or level of education you have completed?

- Less than high school
- High school graduate
- Completed some college
- Associate degree
- Bachelor's degree
- Completed some postgraduate courses
- Master's degree
- Ph.D.
- Prefer not to answer

37. What best describes your current work situation?

- I work 30 hours or more per week
- I work less than 30 hours per week
- I am not currently employed
- Prefer not to answer

38. What is your present occupation?

39. Where do you currently live?

What is your email address?

ANNEX 5: LIST OF PHASE 2 INTERVIEW QUESTIONS IN THE KENYA CASE STUDY

This interview protocol has six parts; Parts I to V are in line with the five dimensions of the CPI Framework and Part VI includes generic demographic questions. A slightly different set of questions (question in Part I to V) was used to interview each interviewee category. The box underneath each question clarifies which interviewees were asked to answer each question. All respondents answered the generic demographic questions in Part VI. Because of the complexity of the topic, for some questions we needed to prompt the interviewee to provide relevant answers, therefore where relevant, these prompts are presented after the questions.

Part I - Goals & Objectives

1. **In your view, what are the most important objectives of Maasai Mara Citizen Observatory?**

   CBM members ☒   GT2.0 team members ☒

2. **To what extent do you think the objectives of Maasai Mara Citizen Observatory are achieved so far?**

   CBM members ☒   GT2.0 team members ☒

3. **If you think there were shortcomings in achieving the objectives, what do you think were the reasons?**

   CBM members ☒   GT2.0 team members ☒

4. **How did you participate in Maasai Mara Citizen Observatory?**

   CBM members ☒   GT2.0 team members ☒

211

5. **Why did you participate in Maasai Mara Citizen Observatory?**

   [instructions] This question is only relevant if the DC lead or team members have participated in the CO beyond the 'regular' project responsibilities (e.g. participation in data collection and sharing).

   CBM members   ☒   GT2.0 team members   ☒

6. **How did you monitor the achievement of the objectives of Maasai Mara Citizen Observatory and the extent of their achievement?**

   CBM members   ☐   GT2.0 team members   ☒

7. **Were there any changes in the objectives of Maasai Mara Citizen Observatory? If yes, what were the changes and why did they happen?**

   [Prompt] Please think of both explicit and implicit (formal/informal) changes in the objectives.

   CBM members   ☐   GT2.0 team members   ☒

Part II - Technology

8. **What is the 'current' architecture of technologies used in Maasai Mara Citizen Observatory?**

   CBM members   ☐   GT2.0 team members   ☒

9. **What were the priority of functionalities and matching tools for Maasai Mara Citizen Observatory?**

   CBM members   ☐   GT2.0 team members   ☒

10. **If there are key differences between the initial and current set of technologies, why did this happen?**

    CBM members   ☐   GT2.0 team members   ☒

11. **Who do you think are the main target users of the tools developed in Maasai Mara Citizen Observatory?**

    CBM members   ☒   GT2.0 team members   ☒

12. **How accessible do you think the tools developed in Maasai Mara Citizen Observatory (e.g. App, website, sensors, test kits, etc.] are for the target users?**

    [Prompt] You can think of accessibility in terms of physical access to technologies or internet or e.g. required skills to use the tools.

    CBM members   ☒   GT2.0 team members   ☒

13. **Which stakeholder groups were involved in designing the functionalities of Maasai Mara Citizen Observatory?**

    CBM members   ☐   GT2.0 team members   ☒

14. **Which stakeholder groups were not involved in designing the functionalities of Maasai Mara Citizen Observatory?**

    CBM members   ☐   GT2.0 team members   ☒

15. **Who are the current end-users of the tools (Apps, web-platforms, etc.) developed in Maasai Mara Citizen Observatory (demographic info, stakeholder category, etc.)?**

    CBM members   ☐   GT2.0 team members   ☒

16. **Which stakeholder groups are currently not using the tools (Apps, web-platforms, etc.) developed in Maasai Mara Citizen Observatory (demographic info, stakeholder category, etc.)?**

    CBM members   ☐   GT2.0 team members   ☒

17. **Do you have any difficulty accessing or using the Maasai Mara Citizen Observatory tools (e.g. App and/or website)? If yes, which one(s) and why?**

CBM members ☒   GT2.0 team members ☐

ii.   <u>Part III - Power dynamics</u>

18. **Since the start of this initiative in March 2017 have you noticed any changes in the social, institutional & political context, which are relevant for the topics of biodiversity conservation and/or livelihood management in the Mara region? If yes, what were the changes?**

CBM members ☒   GT2.0 team members ☒

19. **To what extent, if any, do you think your influence in decision making processes regarding the topics of biodiversity conservation and/or livelihood management in the Mara region has changed because of your participation in Maasai Mara Citizen Observatory? Please explain why?**

[Prompt] Influence can be e.g. via having an official mandate, joining officials for making decisions, providing advice or consult to decision makers, or influencing public opinion.

CBM members ☒   GT2.0 team members ☒

20. **Do you use the data/information produced in Maasai Mara Citizen Observatory? If yes, what do you use this data/information for?**

CBM members ☒   GT2.0 team members ☒

21. **To what extent do you think your access to data/information regarding the topics of biodiversity conservation and/or livelihood management in the Mara region has changed because of your participation in Maasai Mara Citizen Observatory?**

CBM members ☒   GT2.0 team members ☒

22. **To what extent do you think you have control over the data/information that is being produced in Maasai Mara Citizen Observatory?**

    [Prompt] You can think of e.g. access to raw data, or the ability to influence data collection, quality control or data sharing processes.

    CBM members ☒   GT2.0 team members ☒

23. **How was Maasai Mara Citizen Observatory established?**
    - **Top-Down - Maasai Mara Citizen Observatory was initiated and controlled by authorities or stakeholders at higher levels of policy or decision making**
    - **Bottom-Up - Maasai Mara Citizen Observatory was initiated and controlled by stakeholders such as citizens or volunteers who have no official mandate for policy or decision making**
    - **Co-created - Maasai Mara Citizen Observatory was initiated and controlled by a group of interested stakeholders who had the chance to influence its design and functionalities**
    - **Commerce Driven - Maasai Mara Citizen Observatory is a market-based and for-profit initiative**
    - **Other**

    CBM members ☒   GT2.0 team members ☒

24. **What are the envisioned revenue streams for Maasai Mara Citizen Observatory?**

    CBM members ☒   GT2.0 team members ☒

Part IV - Participation

25. **What kind of efforts do you need to put in to participate in Maasai Mara Citizen Observatory?**

    CBM members ☒   GT2.0 team members ☒

26. What kind of support is provided for participation in Maasai Mara Citizen Observatory?

CBM members ☒   GT2.0 team members ☒

27. What are the communication channels in Maasai Mara Citizen Observatory?

[Prompt] Face-to-Face, telephone (call or SMS), email, build-in functionalities on the website, a blog, social media (e.g. Facebook), using apps on your smartphone (e.g. WhatsApp), or any other way.

CBM members ☐   GT2.0 team members ☒

28. To what extent you have so far used the following communication channels in Maasai Mara Citizen Observatory?

	Very frequently	Occasionally	Rarely	Never
Face-to-Face				
Telephone (call)				
Telephone (SMS)				
Email				
Website or Blog				
Social Media (e.g. Facebook, Twitter, etc.)				
Apps on smartphone (e.g. WhatsApp)				
Mass and print media (e.g. radio, TV, etc.)				
Other				

CBM members ☒   GT2.0 team members ☐

Part V - Results

29. What did you expect to see as the outputs (direct products) of Maasai Mara Citizen Observatory?

[prompt] you can think of both positive and negative individual, societal, scientific, economic, environmental and governance-related outputs

CBM members ☒   GT2.0 team members ☒

30. What do you think are the actual outputs (direct products) of Maasai Mara Citizen Observatory?

[prompt] you can think of both positive and negative individual, societal, scientific, economic, environmental and governance-related outputs

CBM members ☒ GT2.0 team members ☒

31. What do you think are the actual short-term or incidental changes (outcomes) that have happened because of Maasai Mara Citizen Observatory?

[prompt] you can think of both positive and negative individual, societal, scientific, economic, environmental and governance-related short-term changes

CBM members ☒ GT2.0 team members ☒

32. What short-term or incidental changes (outcomes) do you expect to see in the near future as a result of Maasai Mara Citizen Observatory?

[prompt] you can think of both positive and negative individual, societal, scientific, economic, environmental and governance-related short-term changes

CBM members ☒ GT2.0 team members ☒

33. What long-term changes (impacts) do you expect to see as a result of Maasai Mara Citizen Observatory?

[prompt] you can think of both positive and negative individual, societal, scientific, economic, environmental and governance-related long-term changes

CBM members ☒ GT2.0 team members ☒

Part VI - Demographic questions

34. What is your age?

- Under 18 years
- 18 to 25 years
- 26 to 35 years
- 36 to 45 years
- 46 to 55 years
- 56 to 65 years
- 66 years or older
- I prefer not to answer

35. What is your gender?

- Male
- Female
- I prefer not to answer

36. What is the highest degree or level of education you have completed?

- Less than high school
- High school graduate
- Completed some college
- Associate degree
- Bachelor's degree
- Completed some postgraduate courses
- Master's degree
- Ph.D.
- Prefer not to answer

37. What best describes your current work situation?

- I work 30 hours or more per week
- I work less than 30 hours per week
- I am not currently employed
- Prefer not to answer

38. What is your present occupation?

39. Where do you currently live?

What is your email address?

ANNEX 6: SCREEN SHOTS OF GRIP OP WATER ALTENA WEB-PLATFORM

(1) Homepage of the Grip op water Altena web-platform

(3) Example of ESRI Survey123 for submitting relevant information on Grip op Water Altena web-platform

(2) Example of Google forms for submitting relevant information on Grip op Water Altena web-platform

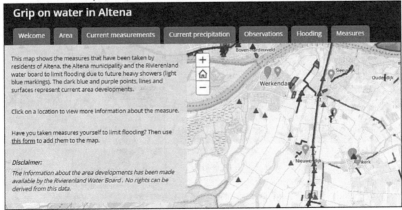

(4) Map of measures taken by different stakeholders to reduce the risk of pluvial flooding

ANNEX 7: SCREEN SHOTS OF MMCO WEB-PLATFORM AND APPS

(1) Homepage of MMCO web-platform

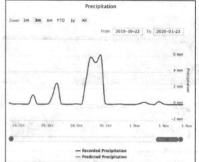

(2) Example of rainfall data from a TAHMO weather station on MMCO web-platform

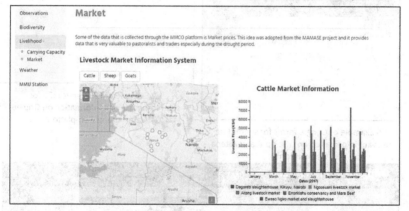

(3) Example of livestock market information on MMCO web-platform

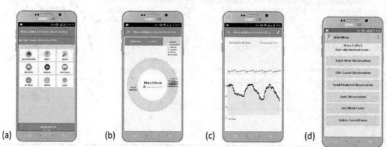

(4) Screen shots of MMCO and Mara Collect Apps: (a) MMCO App - Home page, (b) MMCO App - Chart of submitted observations, (c) MMCO App – weather data, and (d) Mara Collect App- main menu

LIST OF ACRONYMS

AHN	Actueel Hoogtebestand Nederland
ANV	Agrarische Natuur Vereniging
CBM	Community-Based Monitoring
CO	Citizen Observatory
CPI Framework	Conceptual framework for examining the Contextual setting, Process evaluation and Impact assessment of a CBM initiative
DESI	Digital Economy and Society Index
DSA	Daily Subsistence Allowance
ESRI	Environmental Systems Research Institute
EU-FP7	7th Framework Programme for Research and Technological Development in Europe
GIS	Geographic Information System
HFA	Hyogo Framework for Action
ICT	Information and Communication Technology
KFS	The Kenya Forest Service
KNMI	The Dutch Meteorological Institute
KWS	The Kenya Wildlife Service
MARIS	The Mara Rangelands Information System
MMCO	Massai Mara Citizen Observatory
MMU	Massai Mara University
MMWCA	Maasai Mara Wildlife Conservancies Association

NGO	Non-governmental Organization
NSF	US National Science Foundation
ODK	Open Data Kit
OECD	The Organisation for Economic Co-operation and Development
RQ	Research Question
SENSE	Socio-Economic and Natural Sciences of the Environment
SDGs	Sustainable Development Goals
STS	Science and Technology Studies
TAHMO	The Trans-African HydroMeteorological Observatory
UASID	The United States Agency for International Development
UNDP	The United Nations Development Programme
UNECE	The United Nations Economic Commission for Europe
WILD	Wildlife Information Landscape Database
WRMA	Kenya Water Resources Management Authority
ZLTO	The Southern Agriculture and Horticulture Organization

LIST OF TABLES

LIST OF FIGURES

ABOUT THE AUTHOR

 Mohammad Gharesifard was born on 24 February 1981 in Shiraz, Iran. He received his bachelor degree in Civil Engineering from the Islamic Azad University (IAU) in Iran in 2005. He has seven years of combined study/design and construction supervision work experience in Iran's water sector. During this period he worked for consulting companies and was involved in several socio-technical water and sanitation projects. In 2013 he enrolled in Water Resources Management masters program at IHE Delft and graduated with distinction in April 2015. Directly after graduation, Mohammad started a PhD programme under the supervision of Professor Pieter van der Zaag and Dr. Uta Wehn. This research has been funded by the WeSenseIt and Groundtruth2.0 and focuses on understanding the factors that affect the establishment, functioning and outcomes of citizen science projects. Mohammad is interested in studying social, institutional and political aspects of water management practices, Citizen Science, community-based monitoring projects, and public participation processes.

Journal publications

Alonso, S., Mazzoleni, M., Bhamidipati, S., Gharesifard, M., Ridolfi, E., Pandolfo, C., & Alfonso, L. (2020 - in press). Unravelling the influence of human behaviour on reducing casualties during flood evacuation. Hydrological Sciences Journal, Special issue: Advancing socio-hydrology: a synthesis of coupled human–water systems across disciplines.

Gharesifard, M., Wehn, U., & van der Zaag, P. (2019). Context matters: a baseline analysis of contextual realities for two community-based monitoring initiatives of water and environment in Europe and Africa. Journal of Hydrology, 124144. doi:https://doi.org/10.1016/j.jhydrol.2019.124144.

Gharesifard, M., Wehn, U., & van der Zaag, P. (2019). What influences the establishment and functioning of community-based monitoring initiatives of water and environment? A conceptual framework. Journal of Hydrology, 124033. doi:https://doi.org/10.1016/j.jhydrol.2019.124033.

Gharesifard, M., Wehn, U., & van der Zaag, P. (2017). Towards benchmarking citizen observatories: Features and functioning of online amateur weather networks. Journal of Environmental Management, 193, 381-393. doi: http://dx.doi.org/10.1016/j.jenvman.2017.02.003

Gharesifard, M., & Wehn, U. (2016). To share or not to share: Drivers and barriers for sharing data via online amateur weather networks. Journal of Hydrology, Vol. 535, pp.181-190. doi:http://dx.doi.org/10.1016/j.jhydrol.2016.01.036

Book chapters

Gharesifard, M., & Wehn, U. (2016). What Drives Citizens to Engage in ICT-Enabled Citizen Science?: Case Study of Online Amateur Weather Networks, Analyzing the Role of Citizen Science in Modern Research (pp. 62-88). Hershey, PA, USA: IGI Global.

Conference papers, abstracts and workshops

Ajates Gonzalez, R., Wehn, U., Gharesifard, M., Fraisl, D. (2019), Investigating the costs and benefits of citizen observatories in relation to existing in-situ monitoring networks. Presentation at the 11th International Symposium on Digital Earth (ISDE 11), September 24-27, Florence, Italy. [Presenter]

Almomani, A., Awai, E. P., Bonn, A., de Barros, I. A., Friedly, C., Gharesifard, M., … Schade, S. (2019). How does Citizen Science matter for policy? Analyzing the impact of citizen science in policy making. Poster presentation at the iDiv Annual Conference 2019, 29-30 August 2019, Leipzig, Germany. https://doi.org/10.5281/ZENODO.3405334 [Contributor]

Wehn, U., Pfeiffer, E., Anema, K., Joshi, S., Vranckx, S., van der Kwast, H., Giesen, R., Maso, J., Gil-Roldan, E., Ceccaroni, L., Almonani, A., Gharesifard, M., and Giller, O. (2019), Co-designing local knowledge co-production for sustainability: the Ground Truth 2.0 methodology. Assist-UK2019: Association for Studies in Innovation, Science and Technology-UK Annual Conference, University of Manchester, September 9-10, Manchester, UK. [Contributor]

Alfonso, L., Xanthis, A., Mazzoleni, M. Cortes-Arevalo, V.J., Gharesifard M., Wehn u. (2018), Investigating the costs and benefits of citizen observatories in relation to existing in-situ monitoring networks. 2nd International Conference on Citizen Observatories for natural hazards and Water Management, 27-30 November, Venice, Italy. [Presenter]

Shafiei, M., Gharesifard, M., Tavakoli Aminiyan, S., Ghanbari, F., Neyshabouri, S., Davary, K (2018). The current situation and future challenges of Mashhad as a water-wise city in Iran, paper presented at the IWA World Water Congress & Exhibition 2018, 16-23 September 2018, Tokyo, Japan. [Contributor]

Gharesifard, M., Wehn, U., & van der Zaag, P., Masa, A. (2018), Workshop: What can make or break a citizen observatory? The 8th Living Knowledge Conference, 30 May – 1 June 2018, Budapest, Hungary. [Moderator]

Wehn, U., Maso, J., van der Kwast, H., Pfeiffer, E., Giesen, R., Vranckx, S., Pelloquin, C., Cerratto Pargman, T., Gharesifard, M. (2018). The Ground truth 2.0 generic methodology tested in six citizen observatories. PICO presentation at the EGU General Assembly 2018, 8 – 13 April, Vienna, Austria. [Presenter]

Wehn, U., Joshi, S., Pfeiffer, E., Anema, K., Gharesifard, M., & Momani, A. (2017). Addressing the social dimensions of citizen observatories: The Ground Truth 2.0 socio-technical approach for sustainable implementation of citizen observatories. presentation at the EGU General Assembly 2017, 23 – 28 April, Vienna, Austria. [Contributor]

Gharesifard, M., Wehn, U., & van der Zaag, P. (2016). A framework for analyzing the impact of ICT-based citizen science initiatives. Presentation at the International Conference on Citizen Observatories for Water Management, 7 – 10 June, Venice, Italy. [Presenter]

Gharesifard, M., Wehn, U., & van der Zaag, P. (2016). Dimensions of citizen observatories: The case of weather observation networks. Presentation at the 10th GEO European Projects Workshop 2016, 31 May – 2 June, Berlin, Germany. [Contributor]

Gharesifard, M., Wehn, U., & van der Zaag, P. (2016). Dimensions and dynamics of citizen observatories: The case of online amateur weather networks. Presentation at the EGU General Assembly 2016, 17 – 22 April, Vienna, Austria. [Presenter]

Gharesifard, M. and Wehn, U. (2015) Participation in citizen science: Drivers and barriers for sharing personally-collected weather data via web-platforms, Presentation at the 8th International Conference on ICT, Society and Human Beings (ICT2015), 21-23 July, Las Palmas de Gran Canaria, Spain. [Presenter]

Gharesifard, M., & Wehn, U. (2015). Workshop: ICT-enabled Amateur Weather Networks - motivations and barriers for citizen participation, EnviroInfo and ICT4S (pp. 256). Copenhagen, Denmark: University of Copenhagen. [Moderator]

Gharesifard, M., Jahedan, A., & Molazem, B. (2012). Determining the Suitable Sediment extraction Locations of Existing Sand and Gravel Mines on Boshar River in Iran using HEC-RAS Modeling. Paper presented at the 6th International Conference on Scour and Erosion –ICSE-6, Paris, France. [Presenter]

Netherlands Research School for the
Socio-Economic and Natural Sciences of the Environment

D I P L O M A

for specialised PhD training

The Netherlands research school for the
Socio-Economic and Natural Sciences of the Environment
(SENSE) declares that

Mohammad Gharesifard

born on 24 February 1981 in Shiraz, Iran

has successfully fulfilled all requirements of the
educational PhD programme of SENSE.

Delft, 28 September 2020

Chair of the SENSE board

Prof. dr. Martin Wassen

The SENSE Director

Prof. dr. Philipp Pattberg

The SENSE Research School has been accredited by the Royal Netherlands Academy of Arts and Sciences (KNAW)

K O N I N K L I J K E   N E D E R L A N D S E
A K A D E M I E   V A N   W E T E N S C H A P P E N

The SENSE Research School declares that **Mohammad Gharesifard** has successfully fulfilled
all requirements of the educational PhD programme of SENSE with a
work load of 51.5 EC, including the following activities:

SENSE PhD Courses

o   Environmental research in context (2017)
o   Research in context activity: 'Co-organizing UNESCO-IHE PhD Symposium on: From
    water scarcity to water security (Delft, 3-4 October 2016)'

Selection of Other PhD and Advanced MSc Courses

o   The informed researcher: Information and data skills, TU-Delft (2016)
o   The art of presenting science, TU-Delft (2017)
o   Coaching Individual Students and Project Groups, TU-Delft (2017)
o   Preparing for your next career Step in academia, Exploring a career outside academia,
    Personal branding: presenting yourself effectively, TU-Delft (2018)
o   The 8th living Knowledge Conference Summer School, Corvinus university (2018)
o   iDiv Summer School 2019: Citizen Science – Innovation in Open Science, Society and
    Policy, German Centre for Integrative Biodiversity Research (2019)

Selection of Management and Didactic Skills Training

o   Supervising MSc student with thesis entitled 'Understanding flood evacuation at local
    scale using agent-based modelling: a case study in the region of Umbria, Italy' (2018)
o   Lecturing in the MSc course 'Research Skills and Dissemination' (2018-2019)
o   Organisation of an International workshop at 29th EnviroInfo and 3rd ICT4S Conference
    2015 Copenhagen, Denmark
o   Organisation of an International workshop at 8th Living Knowledge Conference 2018,
    Budapest

Oral Presentations

o   *Participation in citizen science: drivers and barriers for sharing personally-collected
    weather data via web-platforms.* 8th International Conference on ICT, Society and
    Human Beings, 21-23 June 2015, 2015, Gran Canaria, Spain
o   *Dimensions and dynamics of citizen observatories: The case of online amateur weather
    networks.* EGU, 17-22 May 2016, Vienna, Austria
o   *A framework for analyzing the impact of ICT-based citizen science initiatives,*
    International conference on Citizen Observatories for Water Management, 7 September
    2016, Venice, Italy
o   *Addressing the social dimensions of citizen observatories: The Ground Truth 2.0 socio-
    technical approach for sustainable implementation of citizen observatories,* EGU, 23-28
    May 2017, Vienna, Austria

SENSE coordinator PhD education

Dr. ir. Peter Vermeulen